Post-Mortem Existence Within the Complex Systems Self-Organization Theory Framework and in Traditional Cultures

Yury N. Kovalyov
National Academy of Culture and Arts Management, Ukraine

Nver M. Mkhitaryan
Institute of Renewable Energy, National Academy of Science of Ukraine, Ukraine & Emerald Palace Group, UAE

Andriy Y. Morozov
Kyiv National University of Trade and Economics, Ukraine

Yaroslava F. Zhukova
State Institution "Soils Protection Institute of Ukraine", Ukraine

Published in the United States of America by
IGI Global
701 E. Chocolate Avenue
Hershey PA, USA 17033
Tel: 717-533-8845
Fax: 717-533-8661
E-mail: cust@igi-global.com
Web site: https://www.igi-global.com

Copyright © 2025 by IGI Global. All rights reserved. No part of this publication may be reproduced, stored or distributed in any form or by any means, electronic or mechanical, including photocopying, without written permission from the publisher.
Product or company names used in this set are for identification purposes only. Inclusion of the names of the products or companies does not indicate a claim of ownership by IGI Global of the trademark or registered trademark.

Library of Congress Cataloging-in-Publication Data

CIP PENDING

ISBN13: 9798369393642
Isbn13Softcover: 9798369393659
EISBN13: 9798369393666

Vice President of Editorial: Melissa Wagner
Managing Editor of Acquisitions: Mikaela Felty
Managing Editor of Book Development: Jocelynn Hessler
Production Manager: Mike Brehm
Cover Design: Phillip Shickler

British Cataloguing in Publication Data
A Cataloguing in Publication record for this book is available from the British Library.

All work contributed to this book is new, previously-unpublished material.
The views expressed in this book are those of the authors, but not necessarily of the publisher.

Table of Contents

Preface v

Introduction xi

Chapter 1
Self-Organization 1

Chapter 2
Post-Mortem Models and Scenarios 33

Chapter 3
Post-Mortem Scenarios Verification Method 67

Chapter 4
Structures Checking 99

Chapter 5
Events Checking: What Old Artifacts and Middle Eastern Texts Can Tell 133

Chapter 6
Events Checking: Indo-Tibetan Version 183

Chapter 7
Details From Low-Ranked Sources 203

Chapter 8
The Transpersonal Experience in Religion and Culture as a Response to the
Challenges of Death 221

Conclusion 251

Compilation of References 263

Related References 275

About the Contributors 309

Index 311

Preface

Since ancient times, humans have sought to understand what to do in the face of inevitable death. Is it still possible to prevent it? The unknown author of the Akkadian version of the Epic tells us about this (George, 2016):

> Shocked by the death of his friend Enkidu, Gilgamesh realizes:
> For his friend Enkidu Gilgamesh
> did bitterly weep as he wandered the wild:
> '1 shall die, and shall 1 not then be as Enkidu?
> Sorrow has entered my heart!
> '1 am afraid of death, so 1 wander the wild,
> to find Uta-napishti, son of Ubar-Tutu.
> On the road, traveling swiftly,
> 1 came one night to a mountain pass.
> IX, 5.

Uta-napishti is a prototype of the biblical Noah, a man who pleased the god Niniguti-Ea and was saved by him during the Deluge, by the will of the gods gained immortality. Thus, Gilgamesh, the great hero, and two-thirds god, one-third man, decides to achieve bodily immortality and for this to meet with Uta-napishti.

Having passed the underworld, he meets Shiduri, the keeper of the Gods tavern, and asks her for the way. Shiduri said (Dyakonov, 2006):

> The tavern-keeper tells him, Gilgamesh:
> Gilgamesh? Where are you aiming?
> The life you seek, you will not find!
> The gods when they created man
> They determined death for a man,
> They held life in their hands.
> You are Gilgamesh, fill your stomach,

> Day and night, may you be cheerful
> Celebrate the holiday every day
> Day and night you play and dance!
> Let your garments be bright,
> Hair is clean, the face is washed,
> Look how the child is holding your hand
> Please your wife with your arms -
> This is the only human destiny!
> X, II.14, III.1-14.

Gilgamesh meets with Uta-napishti, but doesn't understand the quests he is given and is thus forced to return without fulfilling his goal.

After that, he decides to make a name for himself by erecting the walls of Uruk to keep at least the memory of himself immortal - which he completely succeeded.

Our contemporaries see a way out in prolonging the body's life in various means, as well as creating "information copies" of a person, cyborgs, etc. – or at least in a big media name creating in the hope of long-term memory of their achievements.

So, modern humanity is completely within the framework of the Gilgamesh-Shiduri paradigm, without offering any new variants. However, to see new variants, precise knowledge is needed.

Mankind's legacy represents heterogeneous and paradigmatically incompatible knowledge and misconceptions. And, in addition to collecting them, which is a difficult task, they need to be critically assessed, selected, processed, and used to model key life cycle events. And, if the models allow extrapolation beyond the limits of human life, then it is possible to repeat the path of Gilgamesh in its modern sense. And if not, then the question of what lies beyond the threshold of death will remain unanswered until new methods of research are created.

So, the issue of the method is the key. The authors propose to use the theory of complex systems self-organization based on the axiomatic wave model of S-space (Kovalyov, 1996). Its advantage is the ability to model open heterogeneous complex systems in a dissipative environment.

This theory was used previously to model human consciousness and behavior in various historical eras (Kovalyov et al., 2020), as well as to solve applied problems of modeling and optimizing various complex open systems (Mkhitaryan et al., 2004; Shmelova, et al., 2018; Mkhitaryan et al., 2021). The potential of this apparatus for life cycle modeling has been theoretically considered (Kovalyov & Kalashnikova, 2023).

So, models and scenarios for a possible afterlife, are based on a formal mathematical model that describes the evolution and interactions of soliton-wave systems in a dissipative environment. The use of formal methods to explore "eternal" humanitarian questions is rare, but not unprecedented. For example, B. Spinoza used formal apparatus to express his views on God, the mind, etc. (Spinoza, 1677/2020). (Spinoza, 1677/2020).

After a person's death, understood as the destruction of a complex system, its parts continue to exist separately. For example, cells do not die simultaneously, organic matter decomposes within a few weeks, and bones, under favorable conditions, can exist for thousands of years. These facts are well known. But much more interesting is the question of the further existence of human consciousness. Science denies such existence, religion claims the opposite. In this context, various post-mortem scenarios can be predicted, based on life cycle models. Naturally, such scenarios also need to be checked, and it is necessary to find ways of this checking.

Finally, it will be correct to provide information about the modern philosophical understanding of the problem of individual finitude.

Now let us show how this program is implemented in each chapter.

Chapter 1, *Self-Organization*. The *Life Cycle in the Natural Sciences as a Complex System Self-Organization* showed the applicability of various models based on self-organization scenarios. This gives reason to prolong these scenarios for the time after death. For the convenience of readers, we begin the second part with a brief exposition of the S-space wave model axiomatics and the theory of self-organization of complex systems, as well as models of the life cycle and the process of dying.

Chapter 2, *Post-Mortem Models and Scenarios* considers scenarios for the existence of individual components after the destruction of the whole system. They are different: if the existence of a soliton associated with the body goes through further decay (but it can be slowed down if the exchange of substances with the environment is established or special actions are taken), then the existence of a wave associated with the subjective space of a person (the definition is given in (Kovalyov, et al., 2020)) can be extrapolated according to the convolution scenario. This convolution, called the general post-mortem scenario, including stages and special cases, is described in detail and summarized in a table convenient for further verification. The general post-mortem scenario is a predictive scenario for the evolution of subjective space after the collapse of the system, that is, the death of a person.

But how can we test this scenario? After death, medical devices do not record any oscillatory phenomena or electromagnetic fields, and this is a known fact. Appeals to the fact that there are many fields, including those associated with unexplored dark matter and dark energy, lead to the region of the unknown, they cannot be formally refuted at present.

However, there is an indirect version of verification - this is a universal human experience, recorded in numerous artifacts, religious texts, myths, and fairy tales. This is a weak criterion since the sources of such content are not scientific. In addition, the sources are heterogeneous, paradigmatically incompatible, and often difficult to understand. But there is no other way yet, and we use what is available.

Chapter 3, *Post-Mortem Scenarios Verification Method*, substantiates a verification method based on the notions of ternary connection, calibration invariants, and calibrations introduced in Chapter 1. The necessary definitions are introduced, methods of working with sources related to various prehistoric and historical times are considered, 7 criteria for evaluating sources are proposed, and a rating evaluation of sources is made.

In Chapter 4, *Structures Checking*, the levels and channels of human subjective space presented in the *Life Cycle in the Natural Sciences as a Complex System Self-Organization*, are compared with data, obtained from sources. The check showed a good similarity. The structures described in the predicted post-mortem scenarios correspond with mankind's cultural and religious heritage. Levels and channels are also a kind of "coordinate system" that allows us to compare sources with each other.

In Chapter 5, *Events Checking. What Old Artifacts and Middle Eastern Scrolls Can Tell*, the data from Middle Eastern sources about the events that occur for the mental component of a person after the death of the body is compared with the general post-mortem scenario substantiated in Chapter 2 and its cases. The check again shows a good accordance – most sources describe special cases, but there are also descriptions of the general scenario.

In Chapter 6, *Events Checking. Indo-Tibetan Version*, the data from sources of the Indian-Tibetan region are used for a similar comparison. This is done for greater reliability of comparisons since the sources of this group are both geographically and temporally quite distant from the previous ones. Nevertheless, the posthumous events described in them also correlate well with the predictions made in Chapter 2.

Chapter 7, *Details from Low-ranked Sources*, presents various data from low-ranking sources. The verification goals were not set here – the main context of this chapter is interesting details, which are an artistic addition to the main text.

These results, based on the use of the mathematical apparatus, and on sources that were described above as based on non-scientific methods of obtaining knowledge, in turn, need to be comprehended.

Recognizing this, in Chapter 8, *The Transpersonal Experience in Religious and Cultural Practices as a Response to the Challenges of Death*, the authors present a modern philosophical view on the problem. It is philosophy, as a humanitarian science that reflects the entire historical experience of human wisdom, necessary for an adequate comparison with the given data of the natural sciences and mathematical modeling.

The discussion continues in the *Conclusion*, where philosophical views are compared with the results of the work in the form of a dialogue between a mathematician and a philosopher.

The authors hope, that the results of a life cycle analysis will interest the scientific community and a wide readership.

REFERENCES

De Spinoza, B. (2020). *Spinoza's Ethics*. Princeton University Press. (Original work published 1677)

George, A. (2016). *The Epic of Gilgamesh*. Penguin.

Kovalyov, Y., Mkhitaryan, N., & Nitsyn, A. (2020). *Self-organization of the Human Mind and the Transition from Paleolithic to Behavioral Modernity*. IGI Global. DOI: 10.4018/978-1-7998-1706-2

Kovalyov, Y. M., & Kalashnikova, V. V. (2023). Human life cycle modeling. *Suchasni problemy modeljuvannja*. [Modern Problems of Modeling], *25,* 110-122

Shmelova, T., Sikirda, Y., Rizun, N., Salem, A.-B. M., & Kovalyov, Y. (2018). *Socio-Technical Decision Support in Air Navigation Systems: Emerging Research and Opportunities*. IGI Global. DOI: 10.4018/978-1-5225-3108-1

Introduction

Interest in the events of the life cycle has been inherent to people at all times. In one way or another, the culture of all the peoples of the Earth includes fairy tales, myths, epic and literary works, sometimes developed religious ideas and magical rituals, as well as artifacts and monuments reflecting certain events.

However, modern understanding of such issues as the chemical basis, origin, and evolution of life and life cycle, methods of quality and prolongation of life improvement, etc. are associated with the progress of natural sciences and modeling methods.

Among all the stories associated with the life cycle, the theme of post-mortem existence has always stood out: the human consciousness cannot come to terms with the death of loved ones and own death.

At the same time, modern ideas about the life cycle and individual finitude differ significantly from traditional ones.

For traditional ideas, despite their numerous variations and features, two ideas were characteristic:

1. Even if a person's body is mortal, then at least his soul can exist forever;
2. If the soul exists forever, earthly life should provide the best conditions for its posthumous existence.

The modern scientific worldview denies the existence of an eternal soul, which leads to the "everyday" conclusion that – it is necessary to get maximum pleasure from life, trying to prolong the body's (preferably young and healthy) life as much as possible.

However, such a perspective does not satisfy a developed consciousness: a person will never be able to come to terms with the death of loved ones and his own death.

And a person again turns to religion, mythology, fairy tales - or to those humanities sciences, whose subject area is the analysis of such ideas.

The availability of materials is another important feature of our era: everyone is free to choose the "optics" for their research, be it within the framework of religious teaching, personal mystical experience, science, or, for example, works of art. In the same way, within the framework of one paradigm, a person can study, for example, physiological, medical, social, and other aspects of childhood or aging.

The experience of mankind accumulated over thousands of years can rightfully be characterized by the fashionable term - "big data"!

A person, who wants to understand the problems of life and death impartially, falls into an epistemological trap: what to accept as facts and postulates, how to process contradictory and incompatible data, what conclusions to draw, what worldview, and what way of life to consider appropriate?

For such readers, our research will be interesting and useful. Considering, as follows from the above, not only the difference in the subject area but also the large volume and heterogeneity of the materials used, it is advisable to divide it into two parts:

1. Life Cycle in the Natural Sciences as a Complex System Self-Organization.
2. Post-Mortem Existence Within the Complex Systems Self-Organization Theory Framework and in Traditional Cultures.

The description and understanding of evolution are not complete without the evolution of human consciousness and behavior analysis. Therefore, it is worth paying attention to (Kovalyov et al., 2020), dedicated to such an analysis based on the theory of self-organization of complex systems. This book can be considered as a third part of research.

Here we will talk about the second part only.

The results obtained in the *Life Cycle in the Natural Sciences as a Complex System Self-Organization* are the initial data for constructing scenarios of posthumous existence. For the convenience of readers, we begin the second part with a brief exposition of the axiomatics of the wave model of S-space and the theory of self-organization of complex systems, as well as models of the life cycle and the process of dying.

Next, the main scenario of posthumous existence is constructed as a realization of the convolution of the soliton-wave system. With different initial data, some special cases are also possible.

The scenarios allow us to predict the general direction and stages of the process, as well as the activity of the channels and the characteristics of sensations on each of them including personal perception of time. The results are given in Table 4, Chapter 2.

To intrigue the reader, we will not give away the details prematurely!

So, there is a certain theoretical main scenario and several special cases. But how can they be tested when the scientific data has already ended?

The authors have no choice, but to use for verification the richest experience of mankind, expressed in artifacts, monuments, and texts of different eras. The authors realize all the insufficiency and risks of such an approach. In any case, the usage of the proposed mathematical apparatus for these data is interesting because it shows the possibilities and limits of its application.

First, it is necessary to determine the relevance of the sources, and we propose a method for their selection and ranking. A method is based on the construction of a ternary connection and the definition of invariants and calibrations of the sources' content.

Next, a comparison is made of the theoretical structures of the subjective space and the "subtle bodies" and other parts of the "soul" given in various sources. The comparison shows a fairly high similarity between the structures described in the sources, despite the significant differences in terminology and the time of the artifact or text creation.

A similar comparison between the various phases of the contraction and the "events" of the afterlife according to the sources is made. Here, sources from two spatially distant regions are used; in addition, the sources are also separated by time and origin. It is difficult to assess their mutual influence on each other, but the events described fit quite accurately into the main scenario of posthumous existence or its special cases.

Low-rating sources provide a lot of details about the "entrances", "guardians", and "lords" of the world of the dead, their relations, as well as about the "paths" and "topography" of this strange place.

These are colorful and fascinating details, and the authors did not deny themselves the pleasure of talking about them. In addition, this is an opportunity to talk about the views on the afterlife inherent to peoples and cultures with different geographical and historical localizations.

A philosophical generalization is needed to complete this book; the philosopher's position is given both in the last chapter and the conclusion, partly written in a dialogue form.

So, the journey that began in *Life Cycle in the Natural Sciences as a Complex System Self-Organization* continues, to use an expression from the Egyptian Book of the Dead, on a million-year boat! We believe it will be no less fascinating than the excursion to the origins of life in the first part of our research.

The authors hope that their research, methods, and conclusions will be interesting not only to specialists but also to the inquiring reader who makes an effort to gain systematic and comprehensive knowledge about post-mortem existence.

REFERENCES

Kovalyov, Y., Mkhitaryan, N., & Nitsyn, A. (2020). *Self-organization of the Human Mind and the Transition from Paleolithic to Behavioral Modernity*. IGI Global.

Chapter 1
Self-Organization

ABSTRACT

Based on the materials of the Life Cycle in the Natural Sciences as a Complex System Self-Organization, abstractions, and axioms of the wave model of S-space, ternary connections, the structure of human subjective space, and the human-environment model, 4 iterations of the life cycle model are presented. All of them are necessary for developing scenarios of posthumous existence. Abstractions and axioms are needed to understand the basics of the mathematical apparatus used; ternary connections - for the epistemological analysis of texts and artifacts to verify the scenarios; and the structure of human subjective space, the human-environment model, 4 iterations of the life cycle model - for constructing these scenarios.

BACKGROUND

The S-space wave model, created by Y. Kovalyov (Kovalyov, 1996), belongs to mathematics' intuitionistic (constructivist, synthetic) branch (Brauwer, 1907; 1908; Heyting, 1930; 1956; Weyl, 1934). The rejection of the axioms of choice and regularity, as well as the law of exclusion of the third, the replacement of the abstraction of actual infinity by the abstraction of potential feasibility means the rejection of logical formalism (Hilbert, 1950/1902; Hilbert & Bernays, 1934; Aleksandrov, 1956; 1987), which inevitably leads to theoretical multiple paradoxes (Buraly-Forti, Kantor, Russell, etc.).

There is an indirect connection between intuitionistic mathematics and philosophical intuitionism as a method of scientific knowledge in the interpretation of N. Lossky (Lossky, 1993/1904). This connection is important here for solving two problems.

The first is the development of the main method for studying S-space: intuitive design (Kovalyov, 1996; Mkhitaryan et al., 2004).

DOI: 10.4018/979-8-3693-9364-2.ch001

The second is developing a method for interpreting data from various sources used to test some of the predictions of self-organization theory (Chapter 3). This method, called the ternary connective (Kovalyov 1996), is also associated with the philosophical intuitionism of N. Lossky.

Sources on anatomy, physiology (Gaivoronsky et al., 2011; The Human Body, n/d), psychology, (Guyton & Hall, 2005; Ravi Kumar Patil et al., 2009; Schmidt & Tews, 1983), and chronobiology (Frank, 2008; Zidermane, 1988) clearly show that a human being is a complex system. Complex self-organization scenarios are necessary for modeling individual systems, such as the nervous or circulatory systems ({S}, {O}). The general structure of a human is described by a self-organization scenario (1 S, 1 O), based on the deployment of the Fibonacci numerical series.

This scenario is the base of the model of human subjective space and behavior (Kovalyov et al., 2020).

Models of the human-environment interactions, as well as the life cycle and aging process, are described by this scenario. Therefore, it is possible to assume that the scenario of posthumous existence will be its continuation (Chapter 2).

S-SPACE WAVE MODEL

Abstractions

Let us represent Universal (U) which contains everything in the form of an ellipse and divide it into two parts which we call Object (O) and Subject (S). These names have no fixed meaning and may be interpreted differently depending on the situation (as variables x and y in mathematics). As Universal includes everything, hence border between Object and Subject which reflects their differences cannot be introduced from outside, and whereas nothing can be added to Universal, this border doesn't bring in anything new. If we attribute the mark «+» to differences between O and S, differences between S and O must have the opposite mark «-», then the border which contains both differences will be equal to «0», i.e., both conditions will be fulfilled. Let us call *S-space* such a symmetrical border exposed to the influence of S and O and vice versa, which exists since they both exist. The appropriate abstraction of the S-space is a wave.

A wave of S-space "grasps" a particular part of Universal; we will name this part *potential* (π) of S-space and give it a different interpretation depending on the situation. Subjective and objective parts of Sp are symmetrical and it is one of the integrity properties of Sp.

Next, the modalities of actual existence (*A*) is introduced, which means the existing Sp, or S-sets, or S-elements and potential existence (*P*), the possibility of their occurrence (as opposed to non-existence).

Interaction of S and O – *observation* (*H*) – changes the existence modality of Sp and causes its *stratification* (St) into actually existing S- sets and S-elements. An absence of *H* triggers a backward process to potential existing – the *convolution* (Co) of Sp.

S-space (°) is a carrier of sets and elements; *S-sets* (□) – is a potentially or actually existing structure, that coordinates potentially or actually existing and relatively homogenous S-elements.

S-elements include *wave* (∪) – potentially or actually existing element, which is marked by immanent alterability, extension, relative instability, ability to interact, condition modalities, capability to transfer in soliton and *soliton* (●) – a single wave which can transfer in ∪. Entry ∪ may also describe multiple waves; ● – multiple solitons; ∪● – waves and solitons. Elements may have only one component (prototype – a sound wave of one frequency), or several perpendicular components (as an electromagnetic wave that contains perpendicular electric and magnetic elements). Accordingly, single-component, and multi-component waves and soliton waves are distinguished.

AXIOMS

Axiomatic statements of existence 1. 1 -1. 5 (1-st group) codify the ratio of existence modalities (*A, P*) of space, sets, and elements as well as rules of their mutual transfer depending on the dissolution of U and interaction of С и О, i.e., "from up to down".

Axiomatic statement 1.1. *Observation transfers S-space in the condition of actual existence:*

$$S°A \qquad (1.1)$$

Axiomatic statement 1.2. *S-space is a carrier of potentially existing sets that contain potentially existing elements – waves and solitons:*

$$°\supset(\Box P(\cup●P)) \qquad (1.2)$$

Axiomatic statement 1.3. *Observation initiates S-space decomposition into actually existing S- sets:*

$H°P\square A$ (1.3)

Axiomatic statement 1.4. *Observation initiates S-set decomposition into actually existing S-elements:*

$H\square PU\bullet A$ (1,4)

Axiomatic statement 1.5. *Observation initiates S-elements transfer from waves to solitons and vice versa; its stop – the disappearance of elements:*

$HU\bullet A \rightarrow (UA \leftrightarrow \bullet A)$ (1.5)

$U\bullet A \rightarrow U\bullet P\downarrow$

where ↔ – means transfer from waves into solitons and vice versa; ↓ – modality of disappearance.

Axioms 1.2-1.5 form the basis for modeling the processes of self-organization and structuring - as we move from axiom 1.2 to axiom 1.5. the structure of *Sp* becomes more complicated. At the same time, the integrity of the *Sp* is preserved.

The initiator of such self-organization is external influences mentioned in the formulations of all axioms of the first group, which also corresponds to the real conditions of self-organization of living beings.

Opposite phases of the wave are symmetrical; by the influence of external factors symmetry may be broken. When waves are formed during self-organization *Sp*, the symmetry order can increase, and the types of symmetries can also differ.

The presence of solitons and waves, their mutual transitions, and the ability to both develop and degenerate and cease to exist (axiom 1.5) are the basis for explaining the existence of both relatively stable and rapidly changing periods in the existence of individual organisms and ecosystems; the initiation of such transitions depends not only on internal modalities but also on external factors.

The nature of the formation of waves and solitons of the current level based on the elements of the previous one shows how the internal resources necessary for self-organization, self-regulation, and regeneration are used and determine the limits of these processes, which cannot exceed the potential of the elements of the previous level. In addition, a basis is created for explaining why the regeneration of simple organisms is more complete than for complex ones - the potential of *Sp* is limited and can be used either to complicate the structure or to restore; in both cases, external resources are also involved.

Axiomatic statements of existence have top priority.

Axiomatic statement of condition 2.1 (2nd group) defines indicated special aspects of the behavior of the waves. It makes a global link of *Sp*, its sets, and elements on macro and micro levels as a dependence of condition modality of elements from the evolution path of *Sp* and vice versa; dependence of evolution direction of *Sp* from modality condition (æ, ö, ↓) of elements.

Axiomatic statement 2.1. *Stratification progress of S-space defines S- sets structure and condition modality of S- elements – expansion or weakening; condition of elements depicts S- sets the structure and change directivity of Sp:*

$$°P(\Box A((\forall \cup \bullet A(\cup \bullet \emptyset \vee \cup \bullet \ddot{o}))) \tag{1.6}$$

The priority of the axiomatic statement is lower than the axiomatic statement of the first group but higher than the following ones.

The S-space's wave nature, sets, and elements facilitate the modeling of the body-psychic duality. The presence of feedback given by the axiom provides a basis for modeling the reactions of the organism and its effects on the external environment.

Axiomatic statements of interaction 3.1-3.9 (3rd group). It is necessary to distinguish *parallelism* (∥) – the ratio of two or more waves when their interaction doesn't increase the number of components of the resultant wave as compared with initial ones, and *perpendicularity* (⊥) – the ratio of two or more waves when interaction increases the number of components of the resultant wave as compared with initial ones. Entry of the form ∪⊥ means perpendicular waves.

Let's consider the following interactions (operations):

- *Superposition operation* (***S***) – the unification of two or more parallel waves in which components are summed up or compensated.
- *Coincidence operation* (***C***) – the unification of two or more perpendicular S- elements.
- *Interference operation* (***I***) – the interaction of two or more waves and the number of waves, their components, and their condition are changed.
- *Diffraction operation* (***D***) – an interaction of waves and soliton waves and several waves, their condition, and their components are changed (apart from diffraction operation of solitons).

Operations over S- sets, and S-elements are abstractions of different interaction types of physical waves.

Axiomatic statements that describe interaction are applied only with axiomatic statements of existence and condition.

Under superposition operation, parallel components of single or multi-component elements interact. As a result of an operation, their amplitudes are summed up. As axiomatic statement 1.1 is executed, the abovementioned operation doesn't breach conditions of non-binarity and non-additivity.

Axiomatic statement 3.1. *Superposition operation of actually existing single component waves turns them into single component waves with linearly changed properties:*

$$S(\cup A||) \to \cup' \tag{1.7}$$

Axiomatic statement 3.2. *Coincidence operation of actually existing perpendicular S- elements transfers them into the element whose perpendicular components equal to the total amount of perpendicular components of initial waves and soliton waves:*

$$C(\cup \bullet A\perp) \to \cup' \bullet C \tag{1.8}$$

Axiomatic statement 3.3. *Interference operation of actually existing waves turns them into waves with changing a quantity of waves and components and non-linear change of properties:*

$$I(\cup A) \to \cup' \tag{1.9}$$

Axiomatic statement 3.4. *Diffraction operation of actually existing waves and soliton waves transforms waves – into waves with changing the number of waves and components and non-linear change of properties; soliton waves – into soliton waves which retain a set of components and changed movement direction:*

$$D(\cup \bullet A) \to \cup' \tag{1.10}$$

Axiomatic statement 3.5. *There is no such consequence of two and more superposition operations that would turn S-elements into their initial condition:*

$$CC^{-1}(\cup||A) \to \cup'' \tag{1.11}$$

Axiomatic statement 3.6. *There is no such consequence of two or more coincidence operations that would transfer S-elements into their initial state:*

$$CC^{-1}(\cup A\perp) \to \cup''C \tag{1.12}$$

Axiomatic statement 3.7. *There is no such consequence of two or more interference operations that would transform waves into their initial state:*

$$II^{-1}(\cup A) \to \cup' \qquad (1.13)$$

Axiomatic statement 3.8. *There is no such consequence of two or more diffraction operations that would turn waves into their initial state:*

$$DD^{-1}(\cup \bullet A) \to \cup'' \bullet \qquad (1.14)$$

Axiomatic statement 3.9. *There is at least one consequence of two diffraction operations that transforms soliton waves into their initial state:*

$$DD^{-1}(\bullet A) \to \bullet A \qquad (1.15)$$

It is necessary to mention that the properties of soliton waves don't change during diffraction.

Diffraction operation is the most general type of operation in S-space; all others are considered special cases, which emerge depending on the properties of components of interacting elements.

Immanent variability of the components of complex systems from Table 7 of Chapter 1 is realized as the impossibility of inverse operations that transfer S-elements to the initial state, as well as waves and solitons, which can be defined as zero. That's why operations don't make groups except for the special case of diffraction operation of soliton waves (Axioma 3.9).

Axiomatic statements of the measuring 4.1-4.2 (4th group). It is necessary to introduce the following definitions.

A coordinate system (Cs) is called any set of solitons that belongs to different organization levels of *Sp* (layers); due to their interaction with other waves and soliton waves, it's possible to assess their properties. Let's define a total of such layers as *measurable space. The result* (Pc) of measuring should be interpreted as changes in the coordinate system after interaction with a quantifiable element.

The total number of perpendicular components of all soliton waves Cs should be accepted as *the dimensionality* of the coordinate system. *Sp* is an absolute if compared to totals and elements, therefore Cs will describe only measurable subspace. Components Cs are not related to components of the measurable element and depending on its attribute to subspace, there are different grades of comparison of element and Cs.

Measuring is the observation of interaction results of elements and Cs. Such a procedure is not always suitable. Wave and solitons react in different ways when they are measured: the wave is measured till disappearance (axiomatic statement 3.4.), and repeated measurements can't make results more specific; for soliton waves, only movement direction is changed. Cs is also changed.

Axiomatic statements of measuring prescribe interaction procedures with Cs for waves and solitons. They satisfy groups 1-3's axiomatic statements and the special aspects mentioned above.

Axiomatic statement 4.1. *Observation of diffraction operation in the coordinate system and an actually existing wave of measurable space results in changing of coordinate system – measuring result –and changing of the measured wave:*

$$HD(Cs \wedge \cup A) \rightarrow (Cs' \wedge \cup' \wedge Pc) \tag{1.16}$$

Axiomatic statement 4.2. *Observation of diffraction operation of the coordinate system and actually existing soliton in measured space leads to changes in the co-ordinate system – measuring results – and change of direction of soliton:*

$$HD(Cs \wedge \bullet A) \rightarrow (Cs' \wedge \bullet' \wedge Pc) \tag{1.17}$$

Thus, different forms of gaining experience are compared with the fact that the axioms of interactions (except 3.9) do not form groups, as well as with the difference in the behavior of waves and solitons, described by axioms 4.1 and 4.2.

CONSTRUCTION METHODS

Gnoseological concepts are represented by materialistic, idealistic, intuitionistic, and other doctrines. If we abstract from the specific content of the term's "subject", "object", etc., then the process of cognition can be interpreted using projective schemes.

The cognition scheme binarity is derived from the impossibility of cognition without dividing the universe into cognizing and cognizable components – otherwise, there is nothing to cognize and be cognized. Then cognition is an S-O operation:

$$S \leftrightarrow O \tag{1.18}$$

Depending on the orientation, S→O or O→S, is an idealistic or materialistic doctrine. Scheme (1.18) does not express a specific mechanism of cognition; moreover, it is inadequate - if S and O are heterogeneous, then cognition is impossible, and if they are relatively homogeneous, then the scheme does not fix this.

Accounting for boundary operators adds an intra-subject operation of "cognition", which implies binary correspondences between various structures of S. The corresponding scheme looks like this:

$$O \leftrightarrow C(G_c \to G_c^*), \tag{1.19}$$

where G_c and G_c^* are structures of the S (memory and reason), and G is the G-operator.

Compared with (1.18), this scheme expresses relative homogeneity and distinguishes between "cognition" (G-operator) and "reflection" (relation S↔O). This creates a regression ad infinitum. This is proved as follows.

Let the perception of the external world by the senses evoke a reaction in the recognizing system that needs to be recognized. This can be done by recalling the memory and comparing the recognizable reaction with the characteristic features. The resulting reaction again affects the recognition system, which, turning to the memory, must evaluate this new impact, and so on ad infinitum, which contradicts the finiteness of the recognition time.

This implies the inadequacy of (1.19) and the need to supplement it with a structure different from the mind - the nous (Ra), which will be identified with S, as well as the object model (OM). The scheme will take the form:

$$O \to G_c \to G_c^*$$
$$\uparrow \downarrow$$
$$\tag{1.20}$$
$$OM \leftarrow Ra$$

This is the basic binary scheme. We will call (1.20) the scenario of knowledge of the 1st kind, and its product - *knowledge 1 (K1)*.

(1.20) is a formal expression of the rationalistic interpretation of knowledge, which underlies modern methods of studying complex systems. This interpretation faces the following difficulties:

- Identification of "I", S, and Ra contradicts the introspection experience of stopping recursive mental activity, which does not lead to the cessation of cognition.
- The same applies to the phenomena of sleep, loss of consciousness, and subconscious activity, which also do not stop cognition.
- When used for the study of *Sp*, (1.20) does not express the boundaries of *Sp*, does not correspond to the extent and variability of S-sets and S-elements, and does not express the integrity relations of Universe, i.e., is not adequate.

The last condition, also present in (1.18) and (1.19), is the primary source of the difficulties experienced by all binary theories.

The ternary scheme of cognition proposed by A. Bergson and elaborated by N. Lossky (Lossky, 1993/1904):

$$U \rightarrow (O \rightarrow S = \text{«I»})$$

$$\uparrow \downarrow$$

(1.21)

(RN),

where *(RN)* is the general designation of reason and nous; S = "I" - the identification of "I" and S, characteristic of this theory.

Cognition is *"possible because the world is an organic whole, and the cognizer is S, an individual human self, some extra-temporal and extra-spatial being, closely connected with the world"* (Lossky, 1993/1904).

Such an attitude of S to the world - epistemological coordination - is not yet cognition; for cognition *"The subject must direct to the object a whole series of intentional (target) mental acts - awareness, attention, differentiation, etc."* (Lossky, 1993/1904), which is carried out with the help of the sense organs and the recognition system (only RR; memory is understood as a direct contemplation of one's past). The recognition system is potentially tuned to the knowledge of O due to its effects on the senses, but the actual perception occurs only due to the efforts of S (branch S - (RN) - O). Note that, which is important in setting up the recognition system K1, is not yet true knowledge from the point of view of intuitionism.

Cognition occurs when *"a cognized object, even if it is part of the external world, is included directly by the consciousness of the subject, which cognizes, so to speak, into a person and therefore is understood as existing independently of the act of cognition"* (Lossky, 1993/1904) (branch O - S). Such knowledge of O as a result of its connection with S is called intuitive.

The intuitionist theory, however, is not without flaws:

- Highlighted position of "I", like S out of space and time, complicates cognition - if "I" is heterogeneous (RN), then how can S direct target acts? "I" can only be relatively homogeneous (RN). In general, identifying "I" from S contradicts impersonal cognition.
- Interpreting memory only as a contemplation of the past S contradicts the facts of the impermanence of memory data (forgetting, modification during the transition to long-term memory, etc.).
- The possibility of OM adjusting S for the perception of O is not disclosed (and is not considered).

For these reasons, a correction is necessary, after which (3.15) will take the following form:

$$U \to (O, (OM \leftrightarrow (SMNRW \leftarrow «I»)), S), \qquad (1.22)$$

where (SMNRW← "I") means the sense organs, memory, nous, reason, will, and ego of the individual; (OM↔(SMNRW← "I") are interpreted as *Sp*, separating S from O during the decay of U.

Let's explain it in more detail.

1. The process of cognition is understood as the establishment of the Ts of the impersonal S and O in the integral U, where the G operator acts in (SMNRW ← "I"). This is knowledge of the 2ind kind, the product of which will be
   a. impersonal intuitive knowledge - inclusion of O in S in a holistic U;
   b. personal attunement to a given O in the form of an "imprint" of intuitive knowledge in memory and other personality structures.
2. Let us call this personal attunement (2) *knowledge of the 2nd kind (K2)*, distinguishing it from intuitive non-personal knowledge (1) and knowledge K1, which is obtained by logical deriving.
3. Preliminary adjustment is carried out with the help of the target acts of the "I" (which is possible due to the relative homogeneity of the "I" and the SMNRW) or with the help of object models OM of a special type.

Statement 3.1. *K2 is dual, that is, it is both a projection apparatus and a projection, an attunement to O and a model of O.*

Proof.

1. K2, the setting by definition, limits the subjective perception of U to a specific O, which, according to property 3.6., defines K2 as an apparatus of the Ts.
2. Since *Sp* is a boundary space dividing U into subspaces S and O, it is a "plane" in the Universe's space, and K2 belongs to it as a projection of O onto this plane.
3. Since K2 is tuned to a specific O, it contains boundaries that distinguish O from U, and since K2 is not identical to O, it is its model.

It can also be shown that (3.16) expresses (Kovalyov, 1996.):

- Integrity of U.
- Relative homogeneity of *Sp*, S, and O.
- Openness and interdependence of *Sp*, and, consequently, the corresponding characteristics of elements, sets, operations, and mappings.

Consequently, (1.22) *satisfies the necessary conditions and can be used to study S-space directly. The scheme of intuitive design, as the main method of such research, corresponds to (1.22).*

Objectivity and verifiability. First, we consider the existing proof theory to determine the possibilities of known methods. It includes several approaches that are not equally convincing for representatives of different mathematical schools.

1. *The finitistic approach* uses only the abstraction of potential feasibility and extends to constructive objects and functions. Its main advantage is the absence of antinomies and, according to a modern author, *"from a philosophical point of view, it reflects the constructive processes of real nature much more satisfactorily than general set-theoretic mathematics"* (Mathematical Encyclopedia, 1977-85. vol. 2. p. 369). The restrictions correspond to the applicability conditions for K-operators.
2. *The formalization method.* Features and limitations are described above.

3. *Theoretically model methods* are the culmination of formalization; their essence lies in constructing an algebraic system that is a model of a formal theory, provided that all the derived formulas of the latter are true in the semantics generated by the algebraic system.
4. Such modeling is based on the theorems of Gödel (any consistent calculus with classical logic has a model) and Maltsev (if any finite fragment of the calculus has a model, then the whole calculus has a model). In this case, the evidence problem is solved by comparing equivalent models. The restrictions are the same as for the formalization method.
5. *Semi-formal methods.* It is allowed to use an infinite number of constructively generated premises for inference rules. Boundaries - may be somewhat wider than for L- and K-operators.
6. *Non-classical calculus* is not generally accepted. The proof uses more powerful tools than the L and K operators. Judgments are built for known mathematical objects that meet the usual conditions and restrictions.

How objective are these restrictions? If we understand the K-operator only as the T-operator, then they are objective. But this understanding is not the only possible one. The persuasiveness of a proof carried out with the help of effective constructions is not connected with logical proofs; the existence of a "working" model implies certainty, regardless of how it was obtained. Thus, it is possible to use the K-operator to construct an OM as an element of an extra-logical and non-binary Ts (Figure 1).

Figure 1. Interpretation of the intuitive design scheme (1.22)

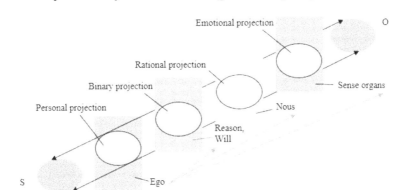

Developing these propositions, we introduce several assertions.
Statement 3.2. *K2 can be expressed in whole or in part using M and OM.*

Proof. Since K2 refers to the structures of the SMNRW← "I", then it is relatively homogeneous M and therefore can be fully or partially expressed by its means. Since K2 is a projection, it can correspond to more than one O; in such a case, O other than the original is the customization object model.

Let us explain the essence of objectivity with the following example. The geometric proof is objective since deduction does not depend on the properties of a particular individual, that is, the impersonal nature of the cognition process ensures objectivity. Fair

Statement 3.3. *K2 is objectively*

Proof. Since intuitive knowledge corresponding to the scheme (1.22) is impersonal, and K2 is its reflection in the SMNRW← "I" structures, obtaining K2 is impersonal, and K2 is objective.

Statement 3.4.*K2 is verifiable.*

Proof. It is enough to indicate the methods of verification - a conclusion on a sufficient basis and OM, which provides practical verification.

The application of the intuitive design method includes the following stages:

- Construction of OM.
- Psychophysiological setting.
- Intuitive perception.
- Fixation and interpretation of the result of perception.

Let's comment on them.

1. The construction of OM is carried out constructively.
2. The need to comply with the boundary conditions gives rise to the following features of the formation of ideas about the object:
   - For tuning, such representations that are not completely rationalistic should be used.
   - View components must be subordinate to the context or composition of the view as a whole.
   - There must be different representations that provide tuning to the same object O.
   - Representations must allow for the impersonal nature of the setting.
3. At the end of perception, it becomes necessary to fix, interpret, and verify K2 when using the resulting OM and M means is possible.

Thus, the intuitive design method, including the construction of constructive OM, the formation of ideas about the object, methods of adjustment and perception, and theoretical and experimental methods for verifying statements, meets the conditions of objectivity and verifiability and can be used for *Sp*.

Statement 3.5. *The intuitive design method, including the building of constructive OM, the formation of ideas about the object, methods of adjustment and perception, and theoretical and experimental methods for verifying statements, meets the requirements of objectivity and verifiability and can be used to study Sp.*

The above reasoning may be used when defining signs, symbols, images, and plots and finding invariants when comparing heterogeneous artifacts and texts created in different eras.

HUMAN-ENVIRONMENT MODEL

Self-Organization Scenario

When modeling relations in the "man-environment" system, we will proceed from the following provisions:

1. The system "human–environment" is considered from the standpoint of the general theory of systems as an integral, complex, open system, which is in a state of dynamic equilibrium. The environment refers to the "close", cognizable world that directly affects the processes of human life.
2. The system is modeled as S-space.
3. The self-organization is modeled as a set of stratification-convolution scenarios.
4. The components, connections, and interactions are modeled S-elements, S-sets, relations, and operations displayed in the OM1 form.
5. The set of stratification-convolution scenarios is determined by the composition of {S} and {O} - "remote" parts of the universe.
6. The problem-solved features determine calibrations.

The "human-environment" model (HEM) can be considered realistic provided that it can describe and make predictions regarding such facts and phenomena:

- Heterogeneity, autonomy, and, at the same time, the integrity of the components of the system (homeostasis of the organism, self-sufficiency of nature).
- The presence of qualitatively different interdependent levels of organization (physiological and mental processes, substance and flight).

- Prevalence of symmetry, constant ϕ, and derivative relations as factors of system integrity.
- Qualitative difference in the interactions of components at different levels - both their coordination and the global nature of the action (systems of regulation, feedback).
- Dismemberment of time scales of processes for different levels - and their mutual consistency.

It follows from the given data on the human body that the most general case of the organization takes place, that is, Sp is formed as a result of the interaction of {S} and {O}.

The general plan of the structure of the human body is simple and clear: the body consists of a symmetrical head and torso; each half of the body has a pair of limbs; each of the limbs consists of three parts (shoulder, forearm, hand and thigh, lower leg, foot, respectively), connected by movable joints and ends with five hoops, consisting of three phalanges, also connected by joints.

This structure is represented by a PC-graph corresponding to scenario (1S, 10), (Figure 2). It is logical to relate the soliton part of the SWM to the body.

Figure 2. The general plan of the structure of the human body

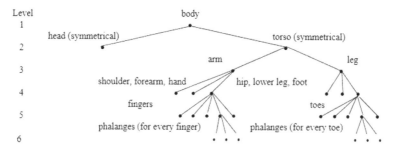

This graph implements the *(1S, 10)* scenario discussed in Chapter 4.

A Human Subjective Space

Now let's see if it is possible to group human sensations similarly. Let us present the facts relating to the physics, physiology, and psychology of perception. The main human senses are vision and hearing.

Let's start with **color perception**. There are seven primary colors. Is there a physical reason why there are seven, or maybe it is due to the structure of the eye or the physiology of perception?

1. *The physical nature of color.* Color depends on the wavelength of light, which is determined by the temperature of the emitting body. It is also necessary to remember the corpuscular nature of light, which determines the discreteness of radiation, depending on which energy level the photons belong to. Since there are many such levels, in reality, we get a certain average length corresponding to a certain body temperature, and we can approximately assume that the series of wavelengths in the part of the spectrum perceived by the eye from violet to red is continuous. Thus, in a series of waves, no peaks or dips separate the ranges corresponding to the primary colors from each other.
2. *Sensitivity of the eye.* In the human eye, about 110-130 million receptors - "rods"- are responsible for the perception of light in general, and 5-7 million "cones" are responsible for the perception of color. The eye cannot perceive a still image; even when we look at one point, the eyes still move, making from 20 to 70 movements per second. With a perfectly transparent atmosphere, retinal rods could react to the light of a candle more than 30 kilometers away. The eye can distinguish 130-250 pure color tones and 5-10 million mixed shades. There is only one perception maximum in the yellow-green color region, not seven, as expected.

Consequently, no physical or physiological prerequisites exist for isolating exactly seven colors. The yellow-green color corresponding to the best perception, is added by ergonomists to the seven basic ones, and thus there are eight.

Interestingly, in various historical eras, the main colors were considered:

- two (dark and light or warm and cold);
- three (preserved in many national flags, in the RGB color rendering system, etc.);
- five (corresponding to the elements or primary elements).

Sound perception. Humans perceive sound non-linearly; there are areas of relatively more and less high sensitivity than the sound pressure level measured using instruments (the first is measured in phons, the second in decibels). The human ear can pick up sounds with a frequency of 10-20 Hz to 15-20 kHz and is most sensitive to the range of 2000 - 2300 hertz, and a person can distinguish 3-4 thousand sounds of different heights. Speech range 1-3 kHz. The best ear for music (the ability to distinguish height) falls in the 80-600 hertz region. Here, the ear can distinguish between sounds with a frequency of 100 and 100.1 hertz.

As with color perception, the recognition of tones as high and low (2 gradations) or high, medium, and low (3 gradations) is in no way connected with either the objective level of continuously changing sound pressure or with the maxima and minima of the sensitivity of the hearing organ. In music, eight fundamental tones (an octave) are distinguished, with a continuous change in pitch.

Other sensory perceptions have been less explored. It is known, for example, that there are about 900 taste buds on the tongue, and their sensitivity depends on temperature.

The area of the olfactory zone of the nose is 5 square centimeters. There are about a million olfactory nerve endings here. For an impulse to arise in the olfactory nerve fiber, about 8 molecules of an odorous substance must get to its end, and for the sensation of smell, at least 40 nerve fibers must be excited.

A human also classifies these sensations regardless of the physical impact and sensitivity of the receptors. Different sources give a different number of tastes, aromas, and other things, which indicate the unformedness of clear sensations.

Sense organs. Receptors perceive hunger and thirst, responsible for balance, coordination, and orientation. They are not grouped into organs but send impulses to the human brain, which classifies them when a certain threshold of sensitivity is reached. The number of receptor sources is determined differently (from 21 to 33).

But there are only five external organs. And this is also inexplicable from the point of view of physics or physiology. Why, for example, sight, hearing, and touch are different senses, if they all perceive external vibrations?

Perception of space. The human eye is a sensitive instrument; there is evidence that it can perceive even single photons. Then, taking into account the difference in the sizes of objects such as:

- Human body – 1.
- Body cells - 10^{-5}.
- Simple molecule - 10^{-9}.
- Atom - 10^{-10}.
- Atomic nucleus - 10^{-14},

one would expect that we see each other not as solid three-dimensional bodies but as separate points of light. But what we see are solid bodies, and this needs an explanation.

On the other hand, physicists do not explain the phenomenon of the three-dimensionality of space. Moreover, if I. Newton considered space as three-dimensional, then A. Einstein's special theory of relativity is defined in a four-dimensional space-time continuum (G. Minkowski's space), and T. Kaluza operated on a five-dimensional space (the addition of the fifth dimension to the four-dimensional continuum made it possible to derive Maxwell's equations of their equations of general relativity). According to E. Noether's theorem, there is a connection between symmetries of various types and conservation laws (Table 1). Modern physicists have calculated that ten-dimensional space (string theory) is needed to implement all known conservation laws, and some theories even consider twenty-six dimensions (bosonic string theory). Therefore, the perception of three-dimensional bodies as integral and three-dimensional in three-dimensional space must be explained.

Table 1. Correspondence of symmetries, invariances, and conservation laws according to the Neter theorem

Transformation	Invariance	Conservation law for
Time broadcast	Uniformity of time	Energy
Space broadcast	Homogeneity of space	Impulse
Rotation of space	Isotropy of space	Moment of impulse
C, P, CP, and T symmetry	Isotropy of time	Parity
Lorenz group	Lorentz covariance	Mass center motion
Calibration conversion	Gauge invariance	Charge

The same applies to the perception of time - we divide it into past, present, and future. The dimensions of space are not completely equal (the third is less accessible due to gravity), so the present is less accessible to us compared to the past and future.

The binarity of judgments (good-bad, yes-no, white-black, etc.) underlies mathematical logic. However, its artificiality, relativity, and inconsistency in many cases with reality were clarified in ancient India (darshans of Nyaya and Vaisheshika), in ancient China (Taoism), and in ancient times (several aporias of Zeno). Also, modern specialists in pattern recognition within the framework of mathematical logic cannot solve, for example, the problem of "turning a tadpole into a frog", but are forced, as in many other cases, to resort to fuzzy logic, introduce categories of feature classes, etc. However, the binarity of assessments and conclusions demonstrates an unusual persistence and needs to be explained.

Unity of truth. A person is inclined to complete his judgments with a single true statement - God is one, the Universe is one, any decision is the only true one, and so on. In most cases, such a reduction is not based on any objective facts and provable judgments but corresponds to our intuitive feeling.

Based on the above, it is possible to model the structure of sensations with a graph (Figure 3). It is logical to correlate the wave part of SWM with the subjective space.

Figure 3. Grouping of human perceptions

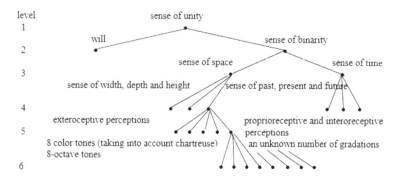

This graph is an implementation of scenario (1S, 1O).

Let's use the graph in Figure 3 to correlate levels and channels with the structures and "abilities" of the human psyche, responsible for the perception of the environment. We will call their totality *the subjective space of a human* - to avoid misunderstandings associated with the concepts of "psyche", "intellect", "mentality", "soul", "and spirit". So:

1. Subjective space is a gestalt conditioned by the general laws of self-organization of complex systems, structuring interactions with the environment.
2. This gestalt includes a set of levels, structures, and channels that are not reducible to each other, each of which has its object of perception.
3. Namely:
    a. The relationship of unity with the environment is intuitively perceived (1channel).
    b. Through the feeling of one's "ego" - the internal unity of the subject (1 channel).
    c. Will and reason affect the object, and the response (2 channels) is evaluated.
    d. The nous operates with spatial and temporal categories (2*3 channels – three dimensions of space, past, present, and future time, grouping sensations as intro-, proprio-, and exteroceptive ones).
    e. Organs of perception determine specific reactions to direct influences (1*5+1*(?)+1*(7) - there are five external sense organs; it is not known exactly how many internal (hunger, thirst, etc.) Sensations exist and how many sensations refers to borderline (for example, a sense of balance). We note once again, that internal and borderline sensations do not have specific organs.

f. Further there is a change in the principle of organization, perceptions become homogeneous, and separate organs of perception are no longer formed (although their rudiments can be seen, for example, in different responses to the light of rods and cones of the human eye). Nevertheless, it is possible further, with some approximation, to trace the correspondence of sensations to the graph in Figure. 2. Thus, the structuring ends here.

4. The graph or PC-diagram of the self-organization scenario *(1S, 1O)* describes the ratio of the channels. The number of channels is defined by the Fibonacci number series - 1-1-2-3-5-8. For example, the mind operates with binary categories according to the "good/bad" principle, the spatiotemporal relations of the mind are ternary, the number of external sense organs is five, etc. Note that the scope of a particular channel covers all subsequent levels, and channels related to subsequent levels inherit one or another of its features since they are formed on its basis. This explains the above quantitative characteristics of the perceptions of external influences by the senses, the three-dimensionality of space and time, and so on. Thus, we state:
    a. consciousness ("ego") manifests at the second and subsequent levels.
    b. reason manifests itself from the third and subsequent levels.
    c. the mind (nous) manifests itself from the fourth level and on subsequent levels.
    d. the fifth level corresponds to the sense organs, and the number of external organs of perception (exteroceptors) is 5.

5. Data about the ability to distinguish shades of sensory sensations are insufficient for unambiguous conclusions. It can be assumed that the quantities of homogeneous sensations are related to Miller's numbers. Independent organs are not formed here (Table 2).

Table 2. Levels and channels of human subjective space

Levels of perception	Objects of perception	Characteristic of perception	Channels, structure, and their number
1	Unity	Humans and the environment are not separate, but the possibility of separation potentially exists.	Intuition 1
2	Separation	Awareness of a person as a self-sufficient whole	+ Ego (mind, consciousness) 1
3	Impact and Reaction	Impact on "not yourself", the response of the environment in response	+ Will (action-reaction) and reason (good-bad) 2

continued on following page

Table 2. Continued

Levels of perception	Objects of perception	Characteristic of perception	Channels, structure, and their number
4	Space and time	Organizing impacts on the environment and its reactions by space and time categories	+ Nous and emotions (tracing forms and changes) 2*3
5	Sensory impacts	Organizing the effects of the environment, internal and borderline sensations by color, sound, etc.	+ Feelings and emotions 2*3*5(?)
6-7	Tones of color, sound, etc.	Distinguishing between shades, tones, etc.	+ Feelings and emotions 2*3*5*8(?)

6. All these channels are present and active constantly, but only one is dominant at some period, which determines the corresponding "state of consciousness". In turn, interaction with a specific object determines the content of the state of consciousness - in the form of an idea, action, word, or emotion, including those related to the past (memory) or future (prediction) time.

7. In psychophysiology, there is no exact understanding of what is meant, for example, by "intuition" - whether it is a sensation, feeling, or ability, whether it corresponds to any physical structure and where such a structure is located, how it relates to "understanding" and "knowledge." Therefore, fundamentally different definitions are offered - by indicating the object of perception, the level of organization, and the characteristics of perception. Thus, each level or channel (S-set or S-element) corresponds to one or another stage of stratification of the S-space and is determined according to Table. 2.

8. We emphasize a certain autonomy of the subjective space both to the physical body of a person (corresponds, except level 6, to the general plan and does not correspond to the structure of the nervous system described by the network graph (6.4)), and the structure of the external environment (more on this below). It is this autonomy, expressed by the sense of one's ego, that explains the emergence of several cognitive and other abilities, for example, the imagination can present images that do not correspond to the environment.

9. An external influence is necessary to start updating the structure (first group axioms), i.e., some metabolic process. The preservation of integrity in this case requires the ratio of the potentials of the levels, which is determined based on the proportion of the golden section. The corresponding weighting coefficients of each level are proportional to the series 1, 0.618, 0.382, 0.236, 0.146, and 0.09. They also determine the contribution of each of the perception channels to holistic image creation.

At the same time, since the potential of the previous level is greater than all the elements of the next one, there is a qualitative difference in the characteristics of both levels (non-additivity), which makes it possible to perceive heterogeneous and different-quality objects. So, the inheritance of the previous level characteristics is due to the transfer of its "qualitatively colored" potential, and the appearance of new features is due to external compensations.

10. We also note the great similarity between a person's mental component and the general structure of his body compared to the animals (examples are considered in (Kovalyov et al., 2020), which allows us to draw the following conclusions:
    a. the balance of the mind and body (integrity) allows a more complex impact on the environment at the level of individuals;
    b. this balance is an evolutionary advantage of man, providing an increase in the likelihood of survival;
    c. striving for the most holistic ratio of body and mentality, expressed by the scenario (1C, 1O), as corresponding to the general laws of self-organization, is the most important factor in evolution, determining, along with environmental conditions, the direction of natural selection.

Human models differ depending on which invariants need to be preserved:

The model in the form of a graph allows you to show the structural correspondence of the general plan of the body structure and subjective space (Figure 4).

Figure 4. Correspondence of the general plan of the body structure and subjective space

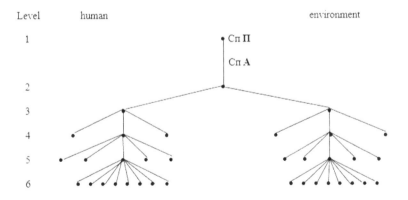

The human body is relatively stable and reacts slowly to environmental influences, while the subjective space is permanently changeable, and its reaction is much faster. In this case, the structure of both bodies and channels does not change.

These characteristics can be shown using the images of soliton and waves. The explanations for why waves are obtained in one case, and solitons in the other, should be sought in the properties and influences of the external environment.

LIFE CYCLE GENERAL MODELS

Modeling of the life cycle based on the relationship of wave and soliton is shown in Figure 4. Both options are suitable for modeling. We will use the option on the left for the image's convenience. Let us show a graph of the change in the value of the *SpA* and *Sp P* potential, which determines the course of the sweep scenario *(1S, 1O)* and its inverse convolution, including the theoretical life span. Recall that the potential is a dimensionless quantity, calibrated depending on the conditions of the problem.

The total potential of *SpA* manifests itself in the formation of the human body and psyche, ensuring the functioning and resistance of the organism throughout life. It is interpreted as an area bounded by the graph of changes in the wave and soliton amplitude values, and the axis shows the lifetime. The current total value of the amplitudes is the potential that can be used at one or another moment of life. We accept the same signature •. Since the variety of factors, and the modeled characteristics are huge we will use the iteration method, moving from simpler models to more complex ones.

Iteration 1 (Figure 5). We will show the theoretical life expectancy and designate the age periods.

Iteration 1 shows a rapid increase of the potential *SpA* during pregnancy and early childhood - it is enough for the formation of the new structure; a decrease in its growth rate to zero in maturity - here it is enough to maintain existing structures and ensure reproduction; a progressive reduction in old age when it is not enough to keep vital parameters and compensate the pathologies.

Iteration 2 (Figure 6). Let's consider the data on the aging of systems and organs, shown in Figure 6, and the actual life expectancy. Unfortunately, these data are not enough to accurately calibrate the model. Death can occur much earlier if one of the vital organs is affected. The same applies to the mental component - a distraught person in natural conditions is doomed to a quick death; in a humane society, of course, he can live much longer.

Following the progressive decrease in Sp A potential, signs of aging are observed: muscle atrophy, reduced growth, and so on. Death occurs due to a such disruption in functioning, which cannot be compensated. And that destroys the integrity of the SWM. The soliton component continues to exist after the destruction of the integral system - this explains the fact that the body does not disappear after death.

Iteration 3 (Figure 7). Let us introduce the next portion of refinements - we will show interactions with the environment. They can be beneficial or destructive. The former increases the potential and reduces it, demanding compensation. In addition, external influences are necessary to construct psychophysiological structures. We also consider and show the following:

- During pregnancy and childhood, the most favorable external conditions are provided, and the potential increases rapidly.
- A person can form useful or bad habits, and fall into favorable or unfavorable conditions, which also affects potential changes.

Figure 5. Life cycle SWM, iteration 1. The value of the potential Sp **A** *is shown conditionally*

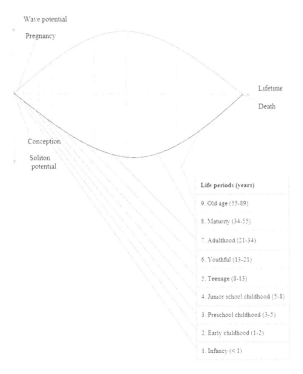

*Figure 6. Life cycle SWM, iteration 2. The minimum potential of the Sp **A**, which required the capacity to maintain functionality is shown in green*

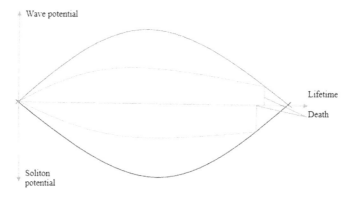

Figure 7. Life cycle SWM, iteration 3. The minimum potential of the SpA, which required the capacity to maintain functionality is shown in green. External influences are shown conditionally and are indicated by contour arrows

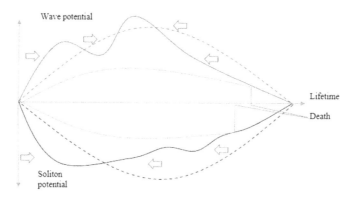

So, due to asymmetric external influences, various asymmetries of the soliton-wave pair arise, which, as noted above, is one of the properties of living systems.

Note that for the mental component, the correct organization of interactions with the environment is necessary - through the intuition, ego, etc channels; powerful intuitive interaction is manifested along the mother-child line.

Iteration 4 (Figure 8). Finally, let's consider the resonance associated with conception. This is the undistributed potential *SpP*, which has not yet expressed itself in forming any structures of the body or psyche but creates the potential possibility of their appearance. It can also be used to compensate for destructive external influences and pathologies. Let us draw a graph of its change separately for the wave and soliton components of the SWM, denoting them as a violet line.

The soliton and the wave already at the moment of gamete fusion have a non-zero potential, and they are consistent with a certain error; after birth, this inconsistency can lead to the destruction of the integrity of the system, which may explain the phenomenon of infant death, which has no apparent cause.

Figure 8. Life cycle SWM, iteration 4. The minimum potential of the SpA, which required the capacity to maintain functionality is shown in green. The purple line shows the change in the SpP potential after the supposed resonance. Arrows indicate external influences

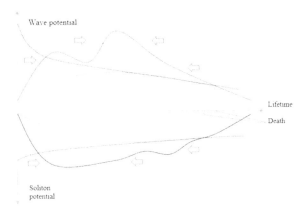

So, SWM iterations 1-4 correspond to the available data on the human life cycle.

FUTURE RESEARCH DIRECTIONS

Directions for further research are constructing a scenario of possible posthumous existence (Chapter 2) and its verification using available sources, including ranking of sources and assessing their compatibility, for which the apparatus of the ternary connection is used (Chapter 3).

CONCLUSION

Abstractions and axioms of S-space are used to study it and create a self-organization theory of open complex systems.

As has already been said, the effectiveness of the study and the reliability of the theory of self-organization are facilitated by the correspondence of axiomatic definitions to the properties of open complex systems.

The ternary connective is one of the verification methods in studying S-space; in the future (Chapter 3) it is supposed to be used to work with various texts and artifacts.

Self-organization scenarios, especially (1S, 1O), are used to model the structure of a person as a whole, his subjective space, connections with the environment, as well as the mechanisms and limits of self-regulation. Since the life cycle is also this scenario implementation, (first - stratification, then convolution), it is supposed to prolong the scenario beyond the event of death. All these material will allow us to construct a scenario of the possible posthumous existence of subjective space, predict its stages, and substantiate the activity of channels and the nature of sensations characteristic of each of them (Chapter 2).

REFERENCES

Anatomy 3D Atlas (n / d). https://anatomy3datlas.com

Brouwer, L. E. J. (1907). *Over de grondslagen der wiskunde*. Amst.-Lpz.

Brouwer, L. E. J. (1908). *De onbetrouwbaarheid der logische principes* (Vol. 2). Tijdsehz voor Wijsbegeerte.

Buzsáki, G. (2007). The structure of consciousness. *Nature*, 446(7133), 267. DOI: 10.1038/446267a

Buzsáki, G., & Watson, B. O. (2012). Brain rhythms and neural syntax: Implications for efficient coding of cognitive content and neuropsychiatric disease. *Dialogues in Clinical Neuroscience*, 14(4), 345–367. DOI: 10.31887/DCNS.2012.14.4/gbuzsaki

Frank, M. G. (2008). *Brain Rhythms*. Frank, M. G. (2008). Brain Rhythms. In *Springer eBooks* (pp. 482–483). DOI: 10.1007/978-3-540-29678-2_727

Gödel, K. (1995). Some basic theorems on the foundations of mathematics and their implications. In *Feferman, ed., 1995. Kurt Gödel Collected Works*, v. III, 304-323. Oxford University Press. (Original work published 1951)

Guyton, A. C., & Hall, J. E. (2005). *Textbook of Medical Physiology* [Textbook]. Saunders. (Original work published 1969)

Heyting, A. (1930). Die formalen Regeln der intuitionistischen Logik. 3 parts, In Sitzungsberichte der preußischen Akademie der Wissenschaften. phys.-math. Klasse. 42-65, 57-71, 158–169.

Heyting, A. (1956). *Intuitionism. An introduction*. North-Holland Publishing Co.

Hilbert, D. (1950). *Grundlagen der Geometrie* [The Foundations of Geometry]. 2nd ed.). (Townsend, E. J., Trans.). Open Court Publishing. (Original work published 1902)

Hilbert, D., & Bernays, P. (1934). *Grundlagen der mathematic* [The Foundations of Mathematics]. Julius Springer.

Kovalyov, Y., Mkhitaryan, N., & Nitsyn, A. (2020). *Self-organization of the human mind and the transition from paleolithic to behavioral modernity*. IGI Global International Publisher.

Ravi Kumar Patil, H. S., Makari, H. K., Gurumurthy, H., & Sowmya, S. V. (2009). *A Textbook of Human Physiology*. I K International Publishing House.

Schmidt, R. F., & Thews, G. (Eds.). (1983). *Human Physiology*. Springer-Verlag.

Tarski, A. (1969). Truth and Proof. *Scientific American* 220: 63–77 (Original work published 1936) The Human Body. (n/d). https://www.healthline.com/human-body-maps

Tononi, G., & Edelman, G. M. (1998). Consciousness and complexity. *Science. 4*(282) (5395),1846-51. DOI: 10.1126/science.282.5395.1846

ADDITIONAL READING:

Aleksandrov, A. D. (1956). *Obshhyj vzghljad na matematyku*. [General view on mathematics]. In *Mathematics, its content, methods, and meaning*. v. 1, 5-78

Aleksandrov, A. D. (1987). *Osnovanyja gheometryy* [Foundations of Geometry]. Nauka.

Kovalyov, Y. N. (1996). *Gheometrycheskoe modelyrovanye erghatycheskykh system: razrabotka apparata* [Geometry Modeling of Man-Machine Systems: The Apparatus Creation]. KMUGA.

Losev, A. F. (1968). *Vvedenye v obshhuju teoryju jazykovykh modelej*. Introduction to the general theory of language models. MGPI.

Lossky, N. O. (1993). *Inuitivism* [Intuitionism]. Progress. (Original work published 1904)

Mathematical encyclopedia (1977-85), vols. 1-5. Sov. Encyclopedia.

Mkhitaryan, N. M., Badeyan, G. V., & Kovalyov, Y. N. (2004). *Erghonomycheskye aspekti slozhnikh system* [Ergonomic Aspects of Complex Systems]. Naukova Dumka.

Shchedrovitsky, P. G. *Kratkoe yzlozhenye ontologhyy SMD-metodologhyy kak systemы pryncypov*. [A concise presentation of the SMD methodology ontology as a system of principles]. *Network Magazine "Centaur"*. Retrieved August 13, 2024, from http://v2.circleplus.ru/archive/s1994/tezis/0/sgatoe/text

Weyl, G. (1934). *Phylosofyja matematyky: sbornyk nauchnykh trudov* [Philosophy of mathematics: a collection of scientific papers]. State Technical and Theoretical Publishing House.

Zidermane, A. A. (1998). *Nekotorye voprosy khronobyologhyy y khronomedycyny: obzor lyteratury* [Some issues of chronobiology and chronomedicine: a literature review]. Zinatne.

KEY TERMS AND DEFINITIONS

The modalities of the state of waves and solitons: *expansion* (æ) – emergence and expansion as they extend and absorb other S- elements and within the cooperation with them; *weakening* (ö) – a decrease of length and destruction within interaction; *disappearance* (↓) – transfer from actual to potential existence. Entries of form ∪ö, •ø mean wave (waves) within the condition of weakening and soliton (solitons) within the expansion.

Parallelism: (∥): the ratio of two or more waves when their interaction doesn't increase the number of components of the resultant wave compared to the initial ones.

Perpendicularity: (⊥): the ratio of two or more waves when interaction increases the number of components of the resultant wave compared to the initial ones. Entry of the form ∪⊥ means perpendicular waves.

Superposition operation: (*S*): the unification of two or more parallel waves in which components are summed up or compensated.

Coincidence operation: (*C*): unifying two or more perpendicular S- elements.

Interference operation: (*I*): the interaction of two or more waves and the number of waves, their components, and their condition are changed.

Diffraction operation: (*D*): an interaction of waves and soliton waves and several waves, their condition, and their components are changed (apart from diffraction operation of solitons).

A coordinate system: (Cs): any set of solitons that belongs to different organization levels of Sp (layers); due to their interaction with other waves and soliton waves, it's possible to assess their properties. A total of such layers is *measurable space.*

The dimensionality of Cs: total number of perpendicular components of all solitons of Cs.

The result of measuring: (Pc): changes in the coordinate system after interaction with a measurable element.

Ternary connection (Ts): connection of the impersonal S and O in the integral U, where the G operator acts in (SMNRW ← "I")

Chapter 2
Post-Mortem Models and Scenarios

ABSTRACT

Death is seen as the destruction of an open emergent system. A model of dying that has been proposed is more accurate than that discussed in Chapter 10 by taking convolution into account. The presence of a more precise model made it possible to consider theoretical scenarios for the separate existence of components after the collapse of the system. The body (substance component) continues to exist after this event; a description of changes after a long time after death is given. The separated existence of the field component is similarly assumed. The scenario of such existence, predicted based on the convolution process, is considered. 5 stages of the post-mortem scenario are described and characterized. The main characteristics are summarized in the Table. 4.

BACKGROUND

To discuss the problem of after-death existence, one should be careful not to go beyond the scientific approach. Therefore, we will try to determine the "optics" through which the study, is as rigorously as possible.

It is necessary to separate theoretical concepts from physical and psychological interpretations. Waves and solitons are abstractions interpreted as field and matter, sensory space and body, but they are not sensory space and body themselves.

Based on the data from Chapter 1, we are prolonging the partial convolution scenarios beyond the soliton-wave system destruction. For this intention to be correct, it is enough to show that: 1) the fact of the destruction of a complex system does not mean the impossibility of the existence of complex systems in general; 2) both soliton and wave components remain complex systems; 3) therefore, the scenarios of their self-organization can be prolonged. Further, one can proceed from

DOI: 10.4018/979-8-3693-9364-2.ch002

the assumption that if these scenarios were adequate for modeling the structures and processes considered in all previous chapters, they are still sufficient for new situations.

Interpretation - sensory space and the body - is another matter. The body does not disappear immediately after death; it passes through several stages and can be traced over time. It can see how well it corresponds to the partial convolution scenario.

But it is no longer possible to verify the existence of subjective space so simply. If, nevertheless, to try to carry out such a check, then, firstly, it is necessary to give the scenario a checkable form, and secondly, to find data for verification and develop a verification method.

More specifically, the scenario must predict changes in structures, processes, and sensations - this is this chapter's task. At the same time, selecting data and developing a test method will already be the task of Chapter 3.

Let us group the sources following the objectives of this research stage.

The use of the S-space wave model to determine the possibility of a separate existence of the system components after it died. This problem is posed and solved for the first time, and relevant literature is absent.

The wave model itself and the theory of self-organization are proposed in (Kovalyov, 1996; 1997). This apparatus will be used to predict, create, and discuss scenarios.

There is an interdisciplinary scientific direction - thanatology - the study of death, which researches the mechanisms of dying and forensic aspects of death, such as bodily changes that accompany death and the postmortem period, as well as wider psychological and social problems related to death (Thanatology, n/d). The scientific literature devoted to the difficulties of thanatology is quite extensive and is presented both as monographs and reference books, for example, (Mannix 2018; Meagher, 2017; Ebenstein, 2017; Thames & Hudson, 2017) and publications in scientific journals (Death Studies; Journal of Historical Sociology; Omega: Journal of Death & Dying; Palliative Medicine). This literature is mainly devoted to the philosophical, social, psychological, and medical problems associated with death, as well as the study of the remains of the body (Wittkowski et al., 2015), and, to a lesser extent, discussions around near-death experiences, which the authors refer to a lifetime (Chapter 1).

The exceptions are publications devoted to the hypothesis of the out-of-body existence of consciousness after death. This hypothesis was put forward in 1982 (Blackmor, 2005); subsequently, she summed up her research as follows: *"It was just over thirty years ago that I had the dramatic out-of-body experience that convinced me of the reality of psychic phenomena and launched me on a crusade to show those closed-minded scientists that consciousness could reach beyond the body and that death was not the end. Just a few years of careful experiments changed all that. I found no psychic phenomena – only wishful thinking, self-deception, experimental error, and, occasionally, fraud. I became a skeptic"* (Blackmore, 2000). Skepticism also prevails in modern research (Lamont, 2007) - the basis for this is the rapid death of brain cells and the lack of brain activity determined by medical devices.

An interesting aspect is the extra somatic nature of human culture, which allows hypotheses about the out-of-body nature of at least some attributes of consciousness (Ivanov, 2013) - this is a modern publication. Still, the founders of this approach are Z. Freud and C. Jung. Psychological and philosophical justifications are put forward here but are not supported by empirical evidence.

It follows, that the methodology for model verifying creation is an important problem; Chapter 12 will be devoted to its consideration.

MODELING

Preliminary Remarks

Recall the main difference between modeling in the framework of the theory of self-organization and the set-theoretic one: the formation of S-sets occurs by single S-space stratification is not a result of combining according to some criterion from individual elements, as for set-theoretic models (Chapter 4).

This happens during the birth of living organisms: the genetic material is not only a program for the development of an individual organism, but also the history of the evolution of its ancestors. This was the starting point when modeling the life cycle (Chapter 1).

But what about the previous steps on Earth and in distant stars? There are also stages in the emergence of the dimensions of space and time, the space and time themselves, some actual, and before that - also the potential integrity of the Universe. They are also the reality of our world. They should be reflected in evolutionary scenarios, as they are presented in the corresponding levels and channels of human interaction with the environment.

Refinement of the man-environment model (Chapter 1). Such an awareness of the similarity of man and the Universe - microcosm and macrocosm - as it was formulated in Antiquity, or their fractality, as it is called in modern mathematics, as well as the mechanism for implementing such similarity, is presented in the model shown in Figure 1.

So, the influence of the environment (blue contour arrows) manifests itself when creating resonance conditions at the conception of a person and further, after the formation of the human body, through the external conditions of its existence. The influence of ancestors (solid blue arrows) corresponds to sexual reproduction. The genetic program that determines human development, self-organization of the environment, and self-organization of a person implement scenario (6.1) and are similar in a similar way.

Refinement of Options for The Soliton-Wave System Destruction

It is necessary to find out exactly where destruction can occur. For the destruction options (soliton, wave components, or both), there are upper limits - where space-time measurements appear (actual destruction), and where a binary division into matter and field appears (potential destruction). For a person, these will be the 4th and 3rd levels; for a body that exists in space-time and is compared with the soliton component, this will be level 4, and for the subjective space, compared with the wave component, it will be level 3. Above these levels such a division is not present, therefore, the possibility of separating these components does not exist.

Figure 1. Considering the first epochs of the evolution of the universe in the PC-graph model of a person

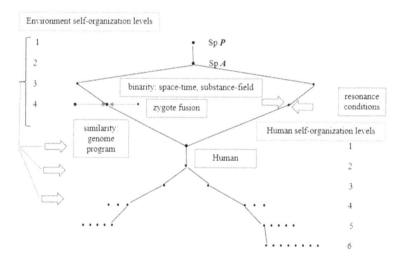

Refinement of the Dying Process

For the destruction of an integral soliton-wave system, the components of which are connected by symmetry relations, conservation laws, and various connections, it is necessary to spend a certain potential. It is essential to distinguish between internal and external factors that form this destruction potential. For example, it can be an injury incompatible with the body's vital functions, caused by an irresistible external force (air crash, shell explosion, etc.), or with the psyche's functions (realization of the meaninglessness of life, emotional shock, etc.). It can also have a relatively weak but long-term impact - poor environmental conditions cause cancer, etc. When the limits of self-regulation are exceeded (Chapter 8), pathology is inevitable. Senility and a decline in strength in the terminal stages can be explained by the expenditure of these forces on self-destruction, for example, cell apoptosis and the like, but an initiating effect is necessary.

Explanation of the "tide of strength" before death. This phenomenon is described in Chapter 10. Within the framework of SWM, it is explained as follows: the potential of the system is higher than the sum of the potentials of its components; in the process of destruction of the system, the potential is "released", which gives a person a certain reserve of strength, both physical and mental, but at the same time serves as a reliable marker of dying.

Death and Convolution

So, considering the clarifications, we will show the options for the soliton-wave system destruction as a PC-graph with convolution considered (Figure 2).

Figure 2. Destruction of the soliton-wave system with convolution considered. 1 - Uncompensated destruction of the wave component; 2 - Uncompensated destruction of the soliton component; 3 - Uncompensated relatively uniform degradation of both components

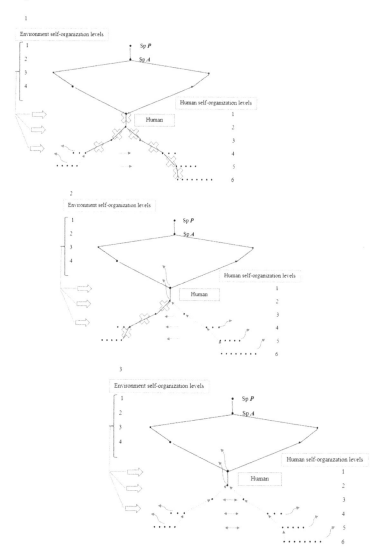

For all scenarios are indicated: blue contour arrows - the influence of the environment; blue solid thin arrows - partial compensation for the loss of potential at different levels; blue dotted arrows - sequences of convolutions; red crosses - destroyed connections and levels; red wavy lines – the scattering of potentials in the corresponding levels of the medium. Thus, in the first scenario, all the structures of both components are destroyed, in the second, all the structures of the soliton (bodily for humans) components, in the third, the structures of the soliton and wave components lose the potential necessary for normal life, but for some time their structures are saved.

PHENOMENA

Now, we will apply the "optics" constructed at the beginning of the chapter and consider how these models correspond to the available data.

First, it is necessary to find out whether the very fact of a person's death that is, the destruction of one complex system, means the impossibility of the existence of complex systems in general. The conditions for the existence of the complex system were formulated in Chapter 1. Strictly speaking, they referred to abiogenic synthesis. Still, they also described the complex open systems' existence conditions, regardless of their nature, and are valid for both soliton and wave components. Table 1 contains only their numbers and names. Let's comment on points 8 and 10.

Table 1. The presence of external conditions for the existence of complex systems after the destruction of the soliton-wave system. Pluses indicate the existence of conditions after the destruction

N°	Condition name	Condition existence
1	Homogeneity of the environment	+
2	Differentiation of the environment	+
3	Symmetry of the environment	+
4	Asymmetry of the environment	+
5	Dissipativity of the environment	+
6	Periodicity of environmental states	+
7	Limits of vital parameters	+
8	External initiating factors	+
9	Possibility of self-organization	+
10	Possibility of selection and inheritance	+

The necessity of some initiating factor for the emergence of life was stated in *Life Cycle in the Natural Sciences as a Complex System Self-Organization* (Chapter 1); initiation is also necessary for the origin of individual organisms, which was discussed in Chapter 6; death is also accompanied by an external impulse for the division of the system, but the potential of the system is released in the process of its division into components. Thus, the initiating factor is present. Also, the fact of system destruction does not negate the possibility of selection and inheritance - the environmental conditions themselves do not change. But without external impulse, it is only a potential possibility, and it doesn't happen. To make it actual, external influence is necessary, as happens in experiments aimed at recreating extinct species since preserved genetic material.

Thus, the soliton-wave system destruction does not change any of the conditions and, therefore, does not prevent the existence of the soliton and wave components as independent complex open systems.

It is also certain, that the soliton component (body) and the wave (sensory space) components are complex systems at a person's death moment and remain so immediately after death.

Now let's see what post-mortem changes occur with the body.

Cadaveric Phenomena

The changes that the organs and tissues of a corpse undergo after the onset of biological death are called cadaveric phenomena (Forensic thanatology, 2011; Cadaveric phenomena, n / d). They are divided into early and late.

Early events include cooling; cadaveric spots; rigor mortis; drying; and autolysis.

Late events include decay; mummification; peat tanning; waxing; skeletonization; and mineralization.

The cooling of the corpse occurs due to the cessation of endogenous heat production, since after biological death, metabolism stops. The first signs of corpse cooling are determined 1-2 hours after the onset of death; after 17–18 hours, the body temperature becomes equal to the ambient temperature.

Cadaveric spots are blue-violet or pale patches of skin. They arise due to the cessation of cardiac activity and the loss of tone of the vascular wall when the blood begins to move through the vessels under the influence of gravity and concentrates in the lower parts of the body during the first 10-12 hours after biological death. The first cadaveric spots may appear after 1–2 hours, and the maximum color intensity is reached by the end of the first half of the day.

Rigor mortis is a contraction of muscle fibers and the following changes. In the striated muscles, signs of rigor mortis are manifested in the form of rigidity and relief. The flexor muscles are more powerful than the extensor muscles, and therefore the upper limbs are flexed at the elbow and hand joints, and the lower limbs are bent at the hip and knee joints. Rigor rigor of smooth muscles is manifested by the so-called "goosebumps", and contraction of the nipples, and sphincters, which leads to the release of excretions. At death, the heart is in diastole. Subsequently, stiffness of the myocardium develops, leading to post-mortem systole and blood extrusion from the heart's ventricles. Because the left half of the heart is more powerful than the right, more blood remains in the right ventricle than in the left. After rigor mortis resolves, the heart returns to diastole. Rigor rigor of the smooth muscles of the gastrointestinal tract forms pronounced folds of the mucous membrane, which can lead to the movement of contents. The rigor of the pregnant uterus, combined with the pressure of abdominal gases, can lead to the expulsion of the fetus ("coffin birth"). Rigor mortis fixes the posthumous posture of the deceased.

Cadaverous desiccation captures mainly those parts of the human body that were moistened during life - the mucous membrane of the lips, the cornea and the whites of the eyes, the scrotum, the labia minora, as well as areas of the skin devoid of the epidermis. The time of appearance and the rate of development largely depends on the state of the environment. After 2–3 hours, under normal conditions, clouding of the corneas is observed, and yellow-brown areas appear on the white membranes of the eye, called "Larcher spots". By the end of the first day, the dried areas become dense to the touch and acquire a yellow-brown or red-brown color.

Cadaveric autolysis is associated with destroying enzyme systems involved in cell metabolism. After the death of a cell, its contents, including intracellular enzymes, fall out of its cytoplasm, and begin to digest everything around; The pancreas and stomach are most susceptible to autolysis. Enzyme systems, spreading uncontrollably, cause rapid disintegration of cellular structures. Cadaveric autolysis occupies an intermediate place between early and late cadaveric changes.

Decay is the decomposition of organic compounds (proteins, fats, carbohydrates, and others) under the influence of microorganisms, forming water, hydrogen sulfide, carbon dioxide, ammonia, methane, and other compounds. After the consumption of oxygen residues by dying cells and bacteria, favorable conditions occur for anaerobic bacteria, especially numerous in the large intestine[1]. They actively consume autolysis products, multiply rapidly, and emit gases. Instead of oxygen, blood hemoglobin attaches sulfur compounds secreted by bacteria and turns into sulfhemoglobin - a dirty green compound that gives the corpse its characteristic color. Thus, the human body's symbionts continue to live and thrive. In the warm season, the corpse is actively populated by insect larvae, especially flies, penetrating the mouth, eyes, and wounds, due to which the corpse begins to lose its biological tissues quickly. Due to the skin's increasing defects, oxygen-loving bacteria are activated again. Complete putrefaction of soft tissues - skin, fatty tissue, muscles, some tissues of organs, and others) Under suitable conditions, it can occur in 3-4 weeks. At the same time, bones, ligaments, cartilage, and formations of many connective tissues are still preserved. When organic compounds are completely decomposed, decay stops.

Mummification refers to the late cadaveric phenomena. For the development of natural mummification, a combination of dry air, good ventilation, and elevated temperature. As a rule, corpses with mild subcutaneous fat and corpses of newborns are subjected to mummification. Many cultures placed great importance on the preservation of the corpse by mummification.

Peat tanning occurs if the corpse is in peaty soil. Under the influence of humic acids, the skin undergoes "tanning", thickens, acquires a brown-brown color, and internal organs decrease in volume. Humic acids contribute to the leaching and dissolving of the mineral basis of bone tissue. At the same time, the bones in their consistency become like cartilage, they are easily cut with a knife. An example of peat tanning is the so-called swamp people (Figure 3).

Figure 3. Head of bog body Tollund Man. Found on 1950-05-06 near Tollund, Silkebjorg, Denmark. The C14 method dated this head to 375-210 BCE. (History of Denmark. DUE, 2006)

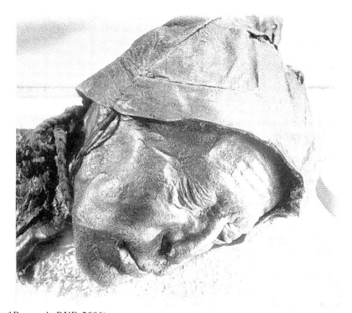

(History of Denmark. DUE, 2006)

Fat waxing (waxing, saponification, saponification) occurs at high humidity and lack of oxygen, most often during burials in moist clay soils when a corpse is in water, and under similar conditions. Under these, putrefactive processes gradually stop, and tissues and organs are saturated with water.

Skeletonization – the disintegration of a corpse into the skeleton's bones; because of decay, soft tissues are destroyed, and ligaments and tendons decompose.

Mineralization – decomposition of a corpse into separate chemical elements and simple chemical compounds. For burials in a wooden coffin and a soil grave, mineralization continues, depending on the soil and climatic conditions, from 10 to 30 years.

At the end of all processes, only the skeleton remains from the corpse, which breaks up into separate bones (teeth, the lower jaw, and diaphyses of long tubular bones are preserved the longest of all hard tissues), and in this form can exist in the soil for hundreds and thousands of years.

There are so-called "corpse farms" - special testing grounds designed for various studies of the processes of decomposition of human bodies, which is important, for example, for forensics (Wolff, 2015). To spare the nerves of readers, instead of photographs from such farms, we use a series of engravings by Eitaku Kobayashi as illustrations, Figure 4.

Figure 4. Eitaku Kobayashi. 9 stages of decomposition of the body of a beautiful courtesan (n / d)

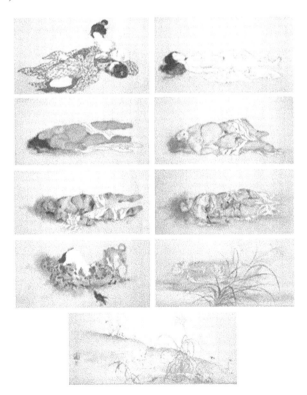

Do these changes match the general convolution scenarios (Chapter 10) and the refined scenarios in Figure 2?

Several important matches can be specified:

- The general direction - simplification and decay - corresponds to scenarios with a decrease in capacity; the initial complexity of the system can no longer be maintained.
- Non-uniformity of the dying systems, organs, and cells of the body (Figure 5) - corresponds to the non-uniformity of the time of existence of solitons belonging to different levels.
- Long-term existence, already in an inanimate form, of parts of the human body - corresponds to partial compensation of potential due to exchanges with the environment.

Figure 5. Stages of decomposition of a corpse. The black line shows the graph of the change in the value of the potential of the soliton component; the blue circle - system transition to another quality

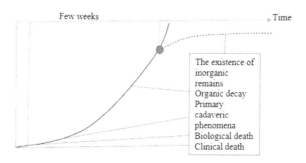

Now we can assume that since both the soliton and the wave components have similar (but not identical convolution scenarios (Figure 2), the changes in the wave component will be similar to post-mortem changes in the body. Let's find out what can be expected from the wave, and what properties should be tested in the future (Table. 2).

Table 2. Preservation of the complex system properties by the • components after the destruction of the soliton-wave system. Designations: + presence of a property; - absence of a property; ? need to check

Properties of complex systems	Stored properties for components:	
	Soliton	Wave
System:	+	+
openness	+	+?
non-additivity	+	+
heterogeneity	+	+?
self-organization (convolution)	-	-?
self-regulation	-	-?
non-linearity of behavior	-	-?
teleological behavior	+	+-?
finiteness		
Components:	+	+?
dependence on the system and each other immanent variability	+	+
Elements:	+	+
immanent extension	+	+
variability	+	+
complexity		
Interactions:	+	+
non-linearity (cases, linearity)	+	+
non-linearity threshold		

Let's explain the entries in the table.

From the complex system properties (Table 8, Chapter 1), the cadaver openness is confirmed, for example, by mummification or mineralization, and can be considered as taking place for subjective space; heterogeneity is evident for both components. The finiteness of the corpse's existence may correspond to the finiteness of the subjective space. Still, it may not correspond - this, like other properties accompanied by a question mark, must somehow be verified since it is unknown for certain about them.

For the components' properties, the variability of both the soliton and the wave components turns out to be obvious; the interdependence of the system components must be checked.

For the elements' properties, the preservation of the remaining characteristics seems to be obvious.

Finally, for interactions, the conservation of system properties also seems obvious.

Thus, if the wave component persists, it is necessary to check first the properties associated with non-additivity and finiteness.

Now let's consider the properties of living organisms formulated in Chapter 2. Let's find out which properties of living organisms are related to the properties of complex open systems (specifically, with the axioms of groups 1-4, Chapter 2), and they need to be checked (Table 3).

Table 3. Verifiable properties that are signs of component life

N°	Property	How expressed for the component:	
		Soliton (body)	**Wave (subjective space)**
1	Isolation from the environment	Homeostasis environment, self-regulation, defensive reactions, etc.	A feeling of one's ego, consciousness, self-preservation instinct, etc.
2	Self-organization and structuring	Systems and organs, cells and organelles, etc.	Levels and channels of ego, will, reason, mind, feelings, etc.
3	External factors	Dependence on the environment, influence on the environment	Dependence on the environment, influence on the environment
4	Symmetries and self-similarity	Axial symmetry, self-similarity of cell groups and cell generations, etc.	Structural symmetry of channels within the level, excitation/inhibition of mental processes, etc.
5	Self-regulation and regeneration	Simple and complex, involving several systems, direct feedback, self-regulation methods, reparation, and regeneration, etc.	Self-control, control of feelings, recovery from emotional experiences, etc.
6	Replication and translation, reproduction and inheritance.	Replication and translation of nucleic acids in cells; reproduction and inheritance of organisms, etc.	Repetition of a sequence of experiences, construction of cause-and-effect chains in reasoning, etc.
7	Variability	Change in height, weight, body structure, etc.	Change in self-awareness, way of thinking, emotional tone, etc.
8	Energy exchange, metabolism	Exchanges with the environment are complex, affecting different types of substances, systems, organs, cells, etc. are involved	The will to influence the environment, assessment of the environment's reactions, making plans, emotional nourishment, etc.
9	Reaction to stimuli	Moving, changing internal parameters, etc.	Reflexes, emotional changes, etc.
10	Periodicity, non-linearity, threshold effects, limitations	Biorhythms, response non-linearity, receptor sensitivity, stress and pathology limits, etc.	Cycles of mental activity, non-linearity of reactions, reaction thresholds, stability and endurance of the psyche, etc.
11	Gaining experience	Natural selection and evolution, etc.	Learnability, etc.
12	Life cycle	An increase in body size and weight at an early age, stabilization in the middle, degradation in later life	Rapid acquisition of knowledge and skills, a complication of mental organization, stabilization, degradation, etc.
13	Death	The finiteness of the existence of the body	The finiteness of the existence of the subjective space

Let's comment on Table 3. First, it is a specification of Table 2 - already to a person. Secondly, its properties are more convenient for verification: for example, the absence of isolation from the environment, self-organization, structuring, self-regulation, regeneration, and so on for a bodily component unequivocally indicates its death. Thirdly, the listed properties apply to systems and to all self-organization levels, which facilitates the process of tracking changes associated with convolution.

We also note that even if it were possible somehow to measure the field activity of the subjective consciousness, this would not determine its content, and, perhaps, would only indicate the direction of the ongoing changes (as under normal conditions, the measurement of the electrical activity of the brain does not indicate the content of thoughts, but speaks, for example, of the transition from relaxation to concentration). Therefore, finding other ways to verify signs of life is necessary.

To do this, we consider the convolution, return to the interpretation of the wave component as a subjective space, and reconstruct the states for different stages of the convolution. We will check their presence in Chapters 4-6.

SCENARIOS

The convolution scenario for the wave part interpreted as subjective space implies several consequences.

The Gradual Change of Consciousness

The main stages correspond to the achieved levels, and transitional stages correspond to the convolution processes. In addition, the initial conditions differ - depending on scenarios 2 or 3, Figure 2. So, let's distinguish (Figure 6):

Here the interaction with the "world of the living" stops - the connection with the material component has already been destroyed and reanimation is impossible.

Figure 6. The gradual change in consciousness

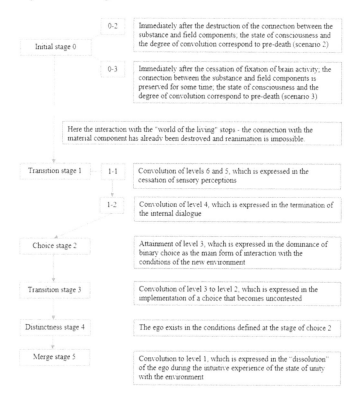

Let's consider each of these stages in more detail following the plan: A. Determine how the potential changes. B. Establish how it is distributed among the levels. C. Define the activity of the channels. D. Establish the nature of sensations; E. Determine what interactions with the environment occur; F) Demonstrate why the subjective space can die. When reconstructing the nature of sensations, we will be guided by (Kovalyov et al., 2020).

Initial Stage 0

A. *Potential changes.* The potential of the subjective part will increase somewhat due to the disintegration of a person, as a complex system, into components; potential will not be spent on maintaining the functioning of the body. These changes continue the death surge effect (Chapter 10).

B. *Distribution of potential by levels.* Based on where the destruction of the integrity of the body and the subjective space of a person occurs (Figure 2), the greatest increase is received by intuition (level 1), ego (level 2), reason and will (level 3), a smaller increase - the mind (level 4) and feelings (levels 5 and 6).
C. *Channel activity.* All channels are active - because of points A and B. In addition, there are effects of liberation from the body with its limitations, and possibly suffering.
D. *The nature of sensations.* Due to point B, the sensations are very vivid. For individual channels, they can be characterized as follows:

> Level 1. Intuition - a vivid sense of unity with external objects, which now includes the deceased body. All other channels and levels continue to function based on intuition, sharing a sense of unity following their properties.
>
> Level 2. Ego - a joyful, euphoric feeling "I'm alive!", Accompanied by bewilderment "Why don't they perceive me?" This feeling of losing one's bodily part creates the basis for "returning" to the body at the stage of clinical death.
>
> Level 3. Will - the possibility of fulfilling desires in the "world of the dead", the lack of the possibility of action in the "world of the living". The reason is a clear binary division of sensations: "I died, not alive", "hell and heaven", "angels and demons", etc. Cognitive dissonance is associated with these circumstances.
>
> Level 4. Nous - the impossibility of ordinary discourse as in the "world of the living", the understanding that in the "world of the dead" ternary divisions of space and time are inadequate, attempts to comprehend the new reality, in particular, the possibilities of free movement (level 2) and sensory sensations (levels 5 and 6).
>
> Levels 5 and 6. Feelings - vivid sensations, free from bodily restrictions. "The world in which I found myself is more real than the one I left" - this is the interpretation given to them by reason.

E. *Interactions with the environment.* All interactions described by the third group axioms (Chapter 2) occur in superposition, coincidence, interference, and diffraction.

However, these are interactions of subjective space with objects of the same quality. There is no interaction with the objects of the "world of the living". This can be explained by the differences in the sizes of various structures noted in Chapter 5 - real objects are perceived as three-dimensional. Still, the wave component "passes through them", and their presence does not affect its state.

F. *Possible causes of death.* In principle, any interaction can destroy a wave (or soliton) corresponding to the subjective space.

So, superposition, if it does not occur at separate levels and not within the limits of self-regulation (Chapter 8), can initiate the "quenching" of the wave" (Figure 7, 1). This requires the coherence of the waves (their identical frequencies and constant phase difference) and the path difference $\Delta s = (k +0.5) * \lambda$, where $k = 0$, $\pm 1, \pm 2, \ldots$, etc.

Also, superposition can lead to the absorption of the original wave (Figure 7, 2). To do this, the following conditions must be met: wave coherence, path difference $\Delta s = k * \lambda$, where $k = 0, \pm 1, \pm 2, \ldots$, etc., the potential of the external wave is greater than the potential of the original one.

Figure 7. Death as a result of superposition: 1 – wave extinction; 2 - absorption by the external wave of the original. Indicated by black lines - waves corresponding to the subjective space; blue lines are the waves interacting with them; orange lines are the results of the interaction. For simplicity, all structures of subjective space are shown as one wave

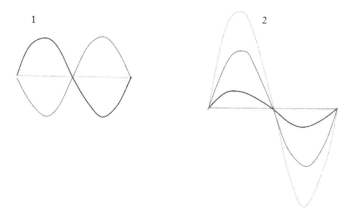

Scheme 1 can be interpreted as the "annihilation of personality". The same scheme may underlie variant 1 of death (Figure 2).

Scheme 2 can be interpreted as "eating the personality" by an external predator.

The coincidence of coherent waves, also under certain conditions corresponding to the polarization (Figure 8), also leads to a violation of the structure of the original wave.

Figure 8. Death due to coincidence. Designations and conventions are the same as in Figure 3

This result can be interpreted as a feeling of "being captured", "connecting with another's body", and the like.

The interference of coherent waves, under certain conditions (Figure 9, shows different options for different initial data), leads to the fact that the potential distribution at different points is significantly different from the initial one for the interacting waves.

Figure 9. Death due to destructive interference

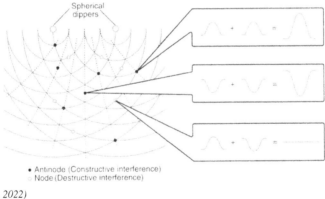

(Veerendra, 2022)

Such an option can be described as "tearing apart" by an external force.

Finally, diffraction, also under certain conditions (Figure 10), also leads to the destruction of waves.

Figure 10. Death as a result of diffraction of waves by solitons

(Dmitris, 2013)

Numerous images of various variants of diffraction and the conditions for their occurrence are given in (Schwarz & Krawczyk, 2020).

This can be interpreted as "breaking apart" from a collision with an external force.

Another option for the death of subjective space at this stage is convolution to level 1, which erases all actual structures formed during the stratification of S-space. This is interpreted as "merging with the light" or "returning to God" - the ideal of adherents of certain doctrines, such as Buddhism or yoga. Rolling up to this stage requires special preparation and conscious exercise.

Transition Stage 1

A. *Potential changes.* The factors contributing to the increase in capacity are no longer in place. The usual sources of its replenishment through the channels of the body have dried up, new ones, through the channels of the subjective space directly, have not yet been established.

   It feels like "hunger" and the reaction is a convolution, first 1-1 and then 1-2.

B. *Distribution of potential by levels.* The potential of levels 5-6 decreases (convolution stage 1-1), and then level 4 (convolution 1-2). Level 3 capacity is increasing, and to a lesser extent, 2.

   This is felt as progressive sensory deprivation (convolution 1-1), poverty and "grayness" of sensations, and "shapelessness" of images. Without emotional nourishment, the internal dialogue gradually stops (convolution 1-2). In turn, the clarity of distinctions also gradually increases, the strength of voluntary impulses (level 3), which causes "movements" from "bad" to "good". At the same time, the "world of the living", which does not give "food" and does not respond to actions, becomes unattractive, "bad", and "movement" comes from it.

C. *Channel activity.* Intuition - reconfigured to new relationships of unity; ego - intensifies; will and reason are strengthened; mind - weakens, but persists; feelings - weaken, but also persist, to a lesser extent than the mind. The catharsis of varying degrees of depth is possible, as well as options with a complete convolution of feelings and mind - more characteristic of individuals striving for such an outcome during their lifetime.

D. *The nature of sensations.* As for the previous stage, let's evaluate the activity of channels by levels.

> Level 1. Intuition - the brightness of sensations is preserved, gradually returning to the perceptions of the "world of the dead" - the communication channels with the "world of the living" are weakened or broken.
> Level 2. Ego – the euphoric feeling of "I'm alive!" gives way to increasing anxiety about safety and finding food, pangs of conscience, fear of "demons", and even panic. At the same time, the interpretation of objects and phenomena of the "world of the dead" is carried out within the framework of a paradigm that was formed during life but is already beginning to change.
> Level 3. This fear generates a "flight" from "bad" to "good": the mind is looking for a way out, whether in a new incarnation, searching for an "angel", crossing the "Styx", trying to hide from a "demon" and the like. The will seeks to carry out the actions prompted by the mind. Those who are indoctrinated, have gained at least relative experience because of various practices in life, or have carried out an appropriate study of the person, have an advantage, representing what is happening.
> Level 4. The nous is "silent": for describing new sensations, the old topoi and discourses are of little use, and the flow of "old" sensations weakens as levels 6 and 5 are reduced.
> Levels 5 and 6. Sensations and feelings weaken, are not interpreted by the mind or not distinguished by the mind, and lose their clarity and effectiveness.

E. *Interactions with the environment.* Theoretically, three group axioms are described (Chapter 2): superposition, coincidence, interference, and diffraction. However, the convolution "loves silence", and interactions harm its smooth passage; it probably happens when they are absent.

F. *Possible causes of death.* All circuits are shown in Figures 5-8 and rolled up to level 1.

Selection Stage 2

This stage is a kind of "testing" of the waves of the subjective space by the "coordinate system" of the waves and solitons of the "world of the dead", which takes place following axioms 4.1, 4.2 (Chapter 2). The result of the interactions will be either the death of the original wave or its inclusion in the composition of the medium - the establishment of "substance exchanges" and so on.

A. *Changes in the potential* at this stage depend on the "setting" of interactions: the potential can be nullified or significantly reduced (Figure 3, scheme 1), or increased (Figure 11), which requires: wave coherence, path difference $\Delta s = k * \lambda$, where $k = 0, \pm 1, \pm 2, \ldots$etc., the potential of the initial wave is greater than the potential of the external one, but within limits that allow self-regulation (Chapter 5).

Figure 11. Absorption by the external wave of the original one. Notation and conventions are the same as in Figure 3

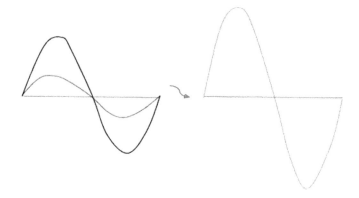

B. *Distribution of potential by levels.* It all depends on the "test results". However, suppose the levels are not destroyed and the original wave does not die. In that case, the distribution remains close to that at the transition stage 1: the reduced potential of levels 6–4 and the increased potential of levels 2–3.
C. *Channel activity.* Intuition is a continuation of restructuring; ego - some stabilization; will and reason - an increase in activity; mind and senses - stabilization.
D. Let's consider *the nature of sensations* by levels.

Level 1. Intuition - the brightness of the sensations of perception of the "world of the dead" is preserved, and the testing situation is perceived as critical, as "tests".

Level 2. Ego - anxiety, pangs of conscience, fear, panic. With a favorable outcome of the "trials" - relief, satisfaction, joy.

Level 3. The mind actively looks for a favorable outcome, selecting criteria based on previous experience or guided by faith or knowledge. The will seeks to carry out the actions prompted by the mind.

Level 4. Nous - continues to be "silent": its way of thinking is inadequate for the new situation, and the potential is weakened by convolution or has already been nullified due to convolution.

Levels 5 and 6. Sensations and feelings are unclear or absent if their potentials are nulled because of convolution.

E. *Interactions with the environment*. All interactions are described by the axioms of the 3-group (Chapter 2): superposition, coincidence, interference, and diffraction. If levels 4-6 are reset, interactions through their channels do not occur.

F. *Possible causes of death*. All circuits are shown in Figures 5-8 and rolled up to level 1.

Transition Stage 2

A. *Potential changes*. Compared to the previous stage, it changes slightly.
B. *Distribution of potential by levels*. It is increasing the potential of levels 1 and 2 due to partial or complete convolution of levels 3-6.
C. *Channel activity*. The rebuilt intuition and ego are active, due to the increase in potential. Passive or degrading will, reason, mind, and feelings.
D. *The nature of sensations* is determined for each of the levels.

Level 1. Intuition - a vague perception of the situation as the final departure from the "world of the living" to forever new conditions.

Level 2. Ego - feeling "I am alive" and submission to circumstances.

Level 3. Reason no longer makes binary distinctions. Space and time are no longer perceived as different. The will weakens and does not attempt to change the situation. Their potential is gradually reset to zero.

Level 4. The mind is "silent": its potential is weakened by convolution or nullified because of convolution.

Levels 5 and 6. Sensations and feelings are unclear or absent if their potentials are nulled due to convolution.

- E. *Interactions with the environment.* All interactions are described by the axioms of the 3-group (Chapter 2): superposition, coincidence, interference, and diffraction. If levels 4-6 are reset, interactions through their channels do not occur.
- F. *Possible causes of death.* All circuits are shown in Figures 5-8 and rolled up to level 1.

Transition Stage 3

- A. *Potential changes.* Compared to the previous stage, it changes slightly.
- B. *Distribution of potential by levels.* It is increasing the potential of levels 1 and 2 due to partial or complete convolution of levels 3-6.
- C. *Channel activity.* The rebuilt intuition and ego are active, due to the increase in potential. Passive or degrading will, reason, mind, and feelings.
- D. *The nature of sensations* is determined for each of the levels.
  - Level 1. Intuition - a vague perception of the situation as the final departure from the "world of the living" to forever new conditions.
  - Level 2. Ego - feeling "I am alive" and submission to circumstances.
  - Level 3. Reason no longer makes binary distinctions. Space and time are no longer perceived as different. The will weakens and does not attempt to change the situation. Their potential is gradually reset to zero.
  - Level 4. The nous is "silent": its potential is weakened by convolution or nullified because of convolution.
  - Levels 5 and 6. Sensations and feelings are unclear or absent if their potentials are nulled due to convolution.
- E. Interactions with the environment. All interactions are described by the axioms of the 3-group (Chapter 2): superposition, coincidence, interference, and diffraction. If levels 3-6 are reset, interactions through their channels do not occur.
- F. Possible causes of death. All circuits are shown in Figures 5-8 and rolled up to level 1.

Distinctness Stage 4

- A. *Potential changes.* Compared to the previous stage, it changes slightly.
- B. *Distribution of potential by levels.* Increasing the potential of levels 1 and 2 by convolving levels 3-6.

C. *Channel activity. Intuition is active.* The ego gradually weakens or stabilizes. Will, reason, mind, and feelings are passive or nullified.
D. *The nature of sensations.* By levels:
> Level 1: Intuition - direct perception of external conditions as infinitely deep "light" or "darkness", depending on the choice at stage 2.
> Level 2. Ego - humility.
> Level 3. Reason does not make binary distinctions. Space and time are perceived as one-dimensional. The will is weak. Their potential is null or has already been nullified.
> Level 4. The mind is "silent": its potential is nullified by convolution.
> Levels 5 and 6. Sensations and feelings are absent; their potentials are eliminated due to convolution.

E. Interactions with the environment. All interactions are described by the axioms of the 3-group (Chapter 2): superposition, coincidence, interference, and diffraction. If levels 3-6 are reset, interactions through their channels do not occur.
F. Possible causes of death. All circuits are shown in Figures 5-8 and rolled up to level 1.

Merge Stage 5

A. *Potential changes.* The potential of the system is reset to zero, "merging" with the potential of the environment.
B. *Distribution of potential by levels.* The potentials of all levels are reset, all structures are no longer supported, and the system ceases to exist as a separate structure.
C. *Channel activity.* Zero for everyone.
D. *The nature of sensations.* By levels:
> Level 1. Intuition - Sensation of merging with the light at the beginning of the stage and no sensation at the end.
> Level 2. Ego - The feeling of final death as a person.
> Level 3. Impulses and estimates are missing.
> Level 4. No thoughts.
> Levels 5 and 6. Sensations and feelings are absent.

E. *Interactions with the environment.* Absorption by the medium, for example, according to schemes 1 or 2 in Figure 5.
F. *Possible causes of death.* Final convolution, modality Sp

Final table. For the convenience of discussion and verification in Chapters 9 and 10, we summarize the most important interpretations described above in Table 4.

Table 4. Reconstruction of sensations in the process of convolution of subjective space

Levels						
1 Intuition	Unity with external objects. All channels and levels continue to function based on intuition, sharing the feeling of unity through their properties	The brightness of sensations returned to the perception of the "world of the dead" - the channels of communication with the "world of the living" are weakened or broken	The brightness of the sensations of perception of the "world of the dead", the testing situation is perceived as "trials"	A vague perception of the situation as a final departure from the "world of the living" to eternally determined conditions	Perception of external conditions as infinitely deep "light" or "darkness", depending on the choice at stage 2	The feeling of merging with the light at the beginning of the stage and the absence of sensation by the end
2 Ego	Euphoric feeling "I'm alive!", accompanied by bewilderment "Why don't they perceive me?"	Interpretation of objects and phenomena of the "world of the dead" within the lifetime paradigm Anxiety, pangs of conscience, fear, panic	Anxiety about safety, nutrition, pangs of conscience, fear. With a favorable outcome of the "trials" - relief, satisfaction, joy	Feeling "I am alive" and submission to circumstances	Humility	The feeling of final death as a person
3 **Reason** **Will**	Clear binary separation of sensations. Possibility to choose between them. The associated cognitive dissonance The possibility of fulfillment of desires in the "world of the dead", the lack of the possibility of actions in the "world of the living"	"Escape" from "bad" to "good": search for a way out (new incarnation, "guide angel", an attempt to hide, etc. The will seeks to carry out the actions prompted by the reason	Active search for a favorable exit; criteria correspond to previous experience, faith, knowledge The will seeks to carry out actions prompted by the reason	Space and time are no longer perceived as different. Will does not attempt to change the situation	Inability to make binary distinctions. Space and time are perceived as one-dimensional Will is weak	No sensations No sensations

continued on following page

Table 4. Continued

Levels						
4 Nous	Impossibility of lifetime discourse, understanding that ternary divisions of space and time are inadequate, comprehension of the new reality: the possibility of free movement, feelings, sensations	"Silence": the old topoi and discourses are unsuitable for describing new sensations, and the flow of "old" sensations weakens as levels are reduced 6 and 5	Continues to be "silent": the way of thinking is inadequate to the new situation; the potential is weakened by the convolution or is nullified because of the convolution	"Silent": the potential is weakened by the convolution or nullified as a result of the convolution	"Silent": the potential is nullified because of the convolution	No thoughts
5, 6 Sensations and feelings	Vibrant sensations, free from bodily restrictions. "The world in which I found myself is more real than the one I left"	Sensations and feelings weaken (perhaps through a sharp surge – of catharsis) and, not interpreted by the mind and not distinguished by the mind, lose their clarity and effectiveness	Sensations and feelings are unclear or absent if their potentials are nulled because of convolution	Sensations and feelings are unclear or absent if their potentials are nulled as a result of convolution	Sensations and feelings are absent, their potentials are nulled because of convolution	Sensations and feelings are absent
Stages	0 - initial	1 - transition	2 - choice	3 - transition	4 - distinctness	5 - merge
Feeling of doom	Annihilation or eating of the personality by an external predator, capture, connection with another's body, disintegration into parts, breaking into parts					Merging with light or darkness at the beginning, no sensation at the end

Let us call the scenario presented in Table 4 the *main post-mortem scenario*.

Let's compare the described picture by changing the potential of the wave component (Figure 12).

Figure 12. Wave component convolution stages. The blue line is a graph of the change in the potential value; the blue circle - system transition to another quality

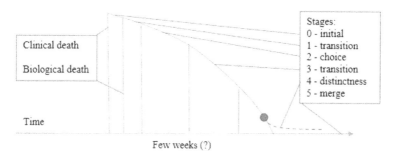

The transition to another quality means that the sensory space "for the living" passes into the sensory space "for the dead".

Following the logic of preserving systems' properties by the soliton-wave system's components after its decay, let us compare Figures 5 and 12. The stage of certainty 4 can be considered analogous to the stage of existence of inorganic remains - in fact, due to exchanges with the environment, a new system is also created here, retaining only some similarity with the old one.

Also, by analogy with the time of transition to this stage for a corpse, we assume that the time to reach the stage of certainty is several weeks - an assumption that needs to be verified. Note that this is "earthly" time; as noted above, already in the early stages of convolution, the sensation of space-time disappears so that this period can be perceived as either very short or very long for subjective space.

Unlike a soliton, the wave part of the SWM is represented by both phases. What can happen to it when it reaches the point of transition into antiphase? The analogy "from the world of waves" suggests 3 options:

- If this place has absorption properties, the wave will be absorbed, and merge stage 5 will come, as shown in Figure 13.
- If the place has " transparency " property, the wave will go into the antiphase (Figure 13.1).
- If the place has the property of "reflection", a standing wave is formed and the transition into antiphase will not occur (Figure 13.2).

Figure 13. Variants of the existence of a wave after reaching the "merge point": 1 – transition into antiphase with the formation of an "anti-soliton-wave pair"; 2 – formation of a standing wave without transition into antiphase with the formation of a soliton-wave pair. Contour arrows show external influences; the wave is blue, and the soliton is black

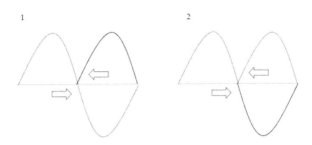

What will the nature of the "anti-soliton-wave pair" be shown in Figure 13.1? From a mathematical point of view, it can be assumed that this will be a system in the Anti-Universe, consisting of anti-matter. Physicists will deny such a possibility; In any case, this is still a purely speculative assumption.

The option shown in Figure 13.2 can be interpreted as follows: a potentially existing wave with nulled actual structures becomes the basis of a new living organism with an external impulse. This zeroing can also explain the effect of "forgetting past lives."

We will call the options shown in Figures 13.1 and 13.2 *reincarnation scenario 1* and *reincarnation scenario 2*, respectively.

For further verification, we will also distinguish two particular scenarios:

- *Resuscitation* - stage 0 is reached, after which the system's integrity is restored because of resuscitation or cessation of trance.
- *Final death* - "absorption" or "torn apart" at any stage, as Figures 7-11 show.

FUTURE RESEARCH DIRECTIONS

The modeling and interpretation of post-mortem states were based on the thesis that if modeling based on the theory of self-organization and the soliton-wave model was adequate for the stages of the emergence and evolution of life, as well as the human life cycle, then it will remain sufficient for modeling after death. Such hope, inspiring at the beginning of the study, is not enough at the end of it - independent verification is needed for the post-mortem scenarios. Developing a verification methodology and its implementation become the immediate problems for Chapters 12-14.

CONCLUSION

The data presented in this chapter and the modeling carried out lead to several conclusions.

1. A person ceases to exist as a holistic complex system after death.
2. Death is not a circumstance prohibiting the separate existence of the substantial and field components as complex systems.
3. And we see that the body (substantial component) does not cease to exist after death, but loses the property of integrity, which, according to the definition given in chapters 2 and 4, means the absence of life. Such signs of integrity as homeostasis and immunity disappear, and decay continues, various cadaveric phenomena are observed, during which some systemic signs are preserved exchanges with the environment.
4. Similarly, we can assume the existence of subjective space (field component) after the death of a person. The laws of self-organization continue to operate, and convolution occurs, prolonging the changes corresponding to aging and due to the progressive insufficiency of the potential to maintain the structures of the subjective space. This assumption is consistent with the results of modeling based on SWM, considering convolution, summarized in Table. 4. Based on the analogy with the time of decay of the flesh, it can be assumed that these changes in subjective space will take several weeks. Here we have existence, but not life, a component, but not a human, and a subjective space, but not a person.
5. The paradox lies in the fact that at levels 1-3 for intuition, ego, will and the reason there is no usual understanding of time, it is "eternity", and even the destruction of subjective space in a finite time from the point of view of an external observer will be subjectively perceived as eternal life, not being such in fact.
6. The end of the convolution is the stage of merging - when the ego ceases to exist - or reincarnation scenarios 1 or 2; in the process of change, a certain stabilization can be achieved, just as swamp people, mummies, or mineralized bones exist for a long time.

We want to emphasize that the data presented do not prohibit the posthumous existence of subjective space, and modeling allows us to describe its scenarios, however, neither the data nor the models are direct evidence. In other words, post-mortem scenarios have the modality of potential but not actual existence. The scenarios follow the logic of mathematical modeling but require independent confirmation.

REFERENCES

Blackmor, S. J. (2000). First person – into the unknown. *New Scientist*, 4, 55.

Blackmor, S. J. (2005). *Beyond The Body: An Investigation of Out-of-the-Body Experiences*. Academy Chicago Publishers.

Dmitris1 (2013). Diffraction of sea waves at breakwater, Ashkelon, Israel. [Illustration]. *File: BreakWaterDiffraction Ashkelon1.jpg - Wikimedia Commons*. (2013, March 24). https://commons.wikimedia.org/wiki/File:BreakWaterDiffraction_Ashkelon1.jpg

Ebenstein, J. (2017). *Death: A Graveside Companion.* Thames & Hudson.

Lamont, P. (2007). Paranormal Belief and the Avowal of Prior Scepticism. *Theory & Psychology*, 17(5), 681–696. DOI: 10.1177/0959354307081624

Kovalyov, Y., Mkhitaryan, N., & Nitsyn, A. (2020). *Self-organization of the Human Mind and the Transition from Paleolithic to Behavioral Modernity*. IGI Global.

Meagher, D. K. (Ed.). (2017). *Handbook of thanatology: The essential body of knowledge for the study of death, dying, and bereavement* (2nd ed.). Routledge.

Puetz, S. J. (2022). The infinitely fractal universe paradigm and consupponibility. *Chaos, Solitons, and Fractals*, 158(112065), 112065. DOI: 10.1016/j.chaos.2022.112065

Schwarz, B., & Krawczyk, C. M. (2020). Coherent diffraction imaging for enhanced fault and fracture network characterization. *Solid Earth*, 11, 1891–1907. DOI: 10.5194/se-11-1891-2020

Veerendra. (2022, November 18). *Analysing interference of waves*. [Illustration]. A Plus Topper; Aplus Topper. https://www.aplustopper.com/analysing-interference-waves/

Mannix, K. (2018). With the End in Mind: Dying, Death, and Wisdom in an Age of Denial. Little, Brown Spark.

Wittkowski, J., Doka, K. J., Neimeyer, R. A., & Vallerga, M. (2015). Publication trends in thanatology: An analysis of leading journals. *Death Studies*, 39(8), 453–462. DOI: 10.1080/07481187.2014.1000054

Wolff, B. M. (2015). A review of "body farm" research facilities across America with a focus on policy and the impacts when dealing with decompositional changes in human remains. [Thesis]. The University of Texas at Arlington. http://hdl.handle.net/10106/25510

ADDITIONAL READING:

Head of bog body Tollund Man. Found on 1950-05-06 near Tollund, Silkebjorg, Denmark, and C14 dated to approximately 375-210 BCE. (2006). [Illustration]. (N.d.). Wikipedia.org. Retrieved August 14, 2024, from https://ru.wikipedia.org/wiki/Болотные_люди

Ivanov, E. M. (2013). Ghypoteza o ekstrasomatycheskoj pryrode pamjaty. [Hypothesis about the extra somatic nature of memory]. *Philosophical Thought*, 8, 1–69. DOI: 10.7256/2306-0174.2013.8.792

Kobayashi. E. 9 stages of decomposition of the body of a beautiful courtesan. [Illustration]. (N.d.). Livejournal.com. Retrieved August 14, 2024, from https://andreas-zarus.livejournal.com/125797.html

Kovalyov, Y. N. (1996). *Gheometrycheskoe modelyrovanye erghatycheskykh system: razrabotka apparata* [Geometry Modeling of Man-Machine Systems: The Apparatus Creation]. KMUGA.

Kovalyov, Y. N. (1997). *Ergonomicheskaya optimizaciya upravleyiya na osnove modeley S-prostranstva* [Ergonomic optimization of control based on S-space models]. KMUGA.

Sudebnaja tanatologhyja. (2011). [Forensic thanatology]. NP IC "YurInfoZdrav"

ENDNOTE

[1] The human body itself is a complex ecosystem inhabited by various parasites, bacteria, and viruses. The body's death does not in the least prevent their continued existence, and since immunity is no longer there, even their prosperity is possible.

Chapter 3
Post-Mortem Scenarios Verification Method

ABSTRACT

The scenarios of the posthumous existence of the subjective space of a person are based on its interpretation as a wave component, which does not disappear after the destruction of the soliton-wave system in the process of dying, just as a dead body does not disappear. This scenario requires careful verification. A methodology for verifying the scenario has been developed. The methodology includes a justification of the verification procedure as an organization of a ternary connection, examples of its application, identification, and ranking of sources as independent information carriers, interpretation of this data, and examples of its use for the ideas from different eras correct comparison.

BACKGROUND

How can we get information about the "world of the dead"? So far, no one has presented credible experimental data to the scientific community.

The near-death experience discussed in *Life Cycle in the Natural Sciences as a Complex System Self-Organization* (Chapter 10) has been recognized as a before-death experience and does not provide the information needed to test a post-mortem scenario.

There is also information about the phenomenon of the ghosts of the dead, collected, for example, by K. Flammarion (Flammarion, 1922). However, they do not meet scientific criteria for validity.

Therefore, looking for information obtained through intuitive perception is necessary - it will inspire confidence among intuitionists. But, according to the data of Chapter 1, it will be expressed either through personal experience (knowledge of the 2nd kind, K2), or through artifacts, symbols, and texts (knowledge of the 1st

DOI: 10.4018/979-8-3693-9364-2.ch003

kind, K1). K2 may be valid for psychologists, and K1 for specialists in semantics and hermeneutics. In turn, both K1 and K2 can be used to build a ternary connective leading to intuitive perception.

Thus, to continue our research, it is necessary to attract data from the humanities. And you should start with an analysis of the concepts of "symbol", "sign", "image", "intention", "meaning", and "plot" from the point of view of the possibility of their use to build a ternary connective. In (Losev, 1993; Lotman, 2000), various definitions and correlations of such concepts are considered.

We single out separate groups of works devoted to the study of symbols (Riera, n / d); Losev, 1982; 1993); images (Bakhtin, 1979; Korshunov, 1991); texts and languages (Losev, 1968; De Saussure, 1977; Van Dyck, 2000); extraction of hidden meanings (Gadamer, 1989); tracking interpretations of symbols in different historical eras (Kovalyov et al., 2020).

Next, one should consider the activity of human perception channels in different historical eras. In (Kovalvov et al., 2020) it is substantiated that the period of dominance of the intuitive channel took place before the emergence of symbolic activity and the formation of developed ideas about the afterlife.

Since the problem of individual finitude has always been at the center of attention of mankind, the peoples of all countries and all continents have many religious teachings and texts, rituals and symbols, myths and fairy tales, as well as other cultural phenomena directly or indirectly related to this topic. It is physically impossible to analyze all these phenomena. Therefore, the problem of choosing sources arises.

First, it is necessary to consider whether the topic of the afterlife is the main one and how thoroughly it is covered. Based on this, the most authoritative are the Pyramid Texts (Mercer, 2020/1952), the Egyptian Book of the Dead (Wallis Budge, 1967/1895), and the Tibetan Book of the Dead (Coleman et al., 2007).

Of great importance for understanding the latter are the Upanishads (Joshi, 2005) - Buddhism, which fundamentally diverged from the Vedic tradition on the issue of sacrifices, nevertheless inherited ideas about the structure of man, the existence of the afterlife, etc. The Tibetan Book of the Dead will be assessed as a source below, and we will return to its comparison with the Upanishads and other texts of the Vedic tradition in Chapters 4, 6.

Ancient Sumerian-Akkadian myths, The Epic of Gilgamesh (Foster, 2001), and Inanna's Descent to the Netherworld (Dedovich, 2020), contain many important details. Still, the theme of the afterlife is not the main one for them.

Nor is it central to the Bible (Holy Bible. The American Standard Version – ASV, n / d). However, the topic is quite fully disclosed in liturgical practice, patristic writings, and modern commentaries relating to the Abrahamic tradition (Vasiliadis, 2012).

Secondly, it is essential to present the sources of different cultures and continents - Africa, Australia, and North and South America.

The death problem is devoted to many works representing the African continent (Abrahamsson, 2009/1951; Paulme, 1967; Mwania, 2016; Baloyi, Makobe-Rabothata, 2018). However, it should be noted that myths, reflecting universal ideas in general and having Neolithic and even Paleolithic roots, differ greatly in detail when moving from region to region. It is impossible to generalize the myths of South and Tropical Africa, the Sahel, the Muslim North, and Christian Ethiopia. Therefore, we will limit ourselves to the references above and not analyze their data in detail.

The mythology of the Australian aborigines is original and sufficiently developed (Berndt & Berndt, 1988/1964). However, the degree of its fixation, as well as the degree of its study, are far inferior to the same Egyptian Book of the Dead. We also will limit ourselves to this reference.

An interesting source is Popol Vuh (Popol Vuh, 2007), a work of relatively late times that has a Christian influence but is valuable in that it expands the geography of sources representing the culture of the Indians of Central America. The Popol Vuh will be used in Chapter 15 as it contains interesting details on defeating the Lords of the Underworld.

The works of M. Eliade (Eliade, 1981-1988) and S. Tokarev (Tokarev, 2005) cite and comment on many examples related to the culture of different eras and peoples, filling in the geographical and cultural diversity of sources.

Less systematized fairy tales and myths of the peoples of the world are also used, given not in primary sources, but in the interpretation of modern authors, which makes it possible to highlight interesting details (Kuhn, 1922; Propp, 1986; Tokarev, 1987; Shirokova, 2004). Among the myths, we single out a group associated with the cult of ancestors (Ermakova, 2002; Keightley, 2004).

All these sources have been evaluated.

VERIFICATION AS A COMPARISON OF INVARIANTS AND THE CONSTRUCTION OF A TERNARY CONNECTIVE

Terms and Definitions

Symbols. The Philosophical Dictionary (The Newest Philosophical Dictionary, 2003) defines: *"A symbol is a concept that fixes the ability of material things, events, sensual images to express ideal contents that are different from their immediate sensual-corporeal being. The symbol has a sign nature, and all the properties of a sign are inherent in it. However, the symbol. turns out to be more than an indication of what he is not. A symbol is not only the name of some particular, it grasps the connection of this particular with many others, subordinating this connection to one law, a single principle, leading them to some single universal. A symbol is* *an independent discovery of reality with its value, in the meaning and power of which, unlike a sign, it participates. Combining various planes of reality into a single whole, a symbol creates his multi-layered structure, a semantic perspective, the explanation and understanding of which requires the interpreter to work with codes of various levels".*

Such definitions are general enough to fix modern ideas about symbols, but they need to be specified for practical application.

According to A. Losev (Losev, 1993), the symbol achieves *"substantial identity of an infinite number of things covered by one model"*, this is the meeting of the signifier and the signified, in which something is identified that, in its immediate content, has nothing in common with each other - symbolizing and symbolized. The symbol has no direct connection and meaningful identity with the symbolized, so similarity does not enter the essence of the symbol.

This understanding, which seems even more abstract, is very close to the understanding of the ternary connective tuning tool expressed by the formula (1.22):

$$U \to (O, (OM \leftrightarrow (SMNRW \leftarrow «I»)), S),$$

where (SMNRW - "I") means the individual's sense organs, memory, nous, reason, will, and ego.

OM is an object model identified with a symbol, considered as Sp, and contains projections of the object, denoting the subject (Chapter 3).

So, in the paradigm of the theory of self-organization, *a symbol is a setting (or part of a setting) of a ternary connective.*

Signs. According to (The Newest Philosophical Dictionary, 2003), *"A sign is a material, sensually perceived object (event, action or phenomenon), acting as an indication, designation or representative of another object, event, action. Designed to acquire, store, convert, and broadcast certain information (messages). Знак is an intersubjective mediator, a structure-mediator in social interactions and communication. Its most important property follows from the definition: being a material object, the sign designates something else. Because of this, the understanding of the sign is impossible without clarifying its meaning - the objective (the object denoted by it); semantic (the image of the designated object); expressive (expressed with the help of his feelings, etc.)".*

Thus, *a sign is a conventional designation with no similarity to the designated object*. It follows that, unlike a symbol, a sign cannot be an element of a ternary connective.

Image. However, signs may well be used to construct an image. The image can be understood in different ways. So, in philosophy, *"An image is the result of the reflection of an object in a person's mind. At the sensory level of cognition, images* *are sensations, perceptions, and representations, at the level of thinking - concepts, judgments, and conclusions. The image is objective in its source - the reflected object, and subjective in the way of its existence. The material forms of the embodiment of the image are actions, language, and various sign models. The image never exhausts all the richness of the properties and relations of the object: the original is richer than its copy, however, having arisen, the image acquires a relatively independent character, and it regulates human behavior.*

In epistemological discourse, the image is directly related to semiotic-linguistic means of expression - from visual signs to conventional signs and symbols in science - and is characterized by the system of interaction between the subject and the object, through the active, transforming attitude of the subject to reality.

The artistic image acts as a way of understanding the world and the language of art. The creative image has many forms: the very process of "subjective deepening" into the material proceeds in a figurative form; images of people, pictures of nature, urban landscapes, images of things, etc. are transformed with the help of imagination, put into new relationships thanks to the constructive thinking of the artist." (Philosophical Encyclopedic Dictionary, 1983).

It follows that the image has the properties necessary for building a ternary connective; its role will be like that of a symbol. However, unlike a symbol, an image causes a more emotional reaction (level 5), it is more closely connected with the author's imagination, and, therefore, is less "accurate" and can only be used to attract attention and initial adjustment.

Intention. When working on a work, the author determines the goal - to convey or hide some knowledge, evoke certain emotions, manipulate the behavior of the reader or viewer, and so on. Further, depending on this, he selects expressive means that communicate with the reader, forming the author's intention. The idea has the necessary characteristics and can be used to construct a ternary connective. In this sense, it is like the projected image, however, on a conditional scale between a sign and a symbol, it will be closer to the sign - *"a thought uttered is a lie"* (F. Tyutchev, Silentium).

Meaning. The created work begins to live its own life. It can be perceived completely differently than the author intended, especially a long time after its creation when the cultural context is forgotten, and the "spirit of the era" disappears. Thus, each reader puts his meaning into the text. The variability of interpretation brings the meaning closer to the symbol, which is a positive characteristic for building a ternary connective. However, the accuracy of such tuning also varies widely, and this feature necessitates careful use of meanings. Note that both the idea and the meaning, unlike the image, are still more rational, and belong to level 4 rather than 5.

The **plot** is a sequence of interrelated events that reveal the author's intention or the meaning of the work. From the point of view of building a ternary connective, the plot gives the idea and meaning more "depth" and more "accuracy", bringing their role closer to that of a symbol. The plot corresponds to level 4 and, to a lesser extent, 5. For a painting, the plot corresponds to the image's subject, image, or symbol.

Having defined these key concepts, let's proceed to the description of the verification sequence.

The Procedure and Examples

Artifacts, symbols, rituals, texts, myths, fairy tales, etc. differ significantly. Therefore, it is necessary to determine what exactly will be compared.

From the standpoint of the theory of self-organization, this can be expressed in terms of "invariants" and "calibrations".

An *invariant* is something common to all the main sources.

Calibration is one detail that distinguishes the data of one of the main sources from others, or individual information from myths and tales that fit into the invariant framework.

So, when mentioning hell and paradise, the invariants will be information about:

- The final place of existence of the soul.
- The eternity of this place.
- Binary gradation of place.
- The eternity of the soul itself.
- Binary attributes of the soul (sinfulness, righteousness) determine the place of its existence (hell or heaven).

This example's calibration will include details about the location of hell and heaven, the number of Houris in heaven, the types of torture in hell, and others.

Accordingly, verification materials must be prepared to determine their invariants and calibrations. The invariants must be compared with data from Table 4 in Chapter 2.

Comparison of materials from different eras. In terms of the theory of self-organization (Chapter 3), this is the problem "about the possibility of constructing a ternary connective" (3.16) - if the symbols allow you to form a connective "directed" to the same invariant, then the texts containing them can be correctly compared. This requires that the symbols:

1. Acted on all or most channels of the human sensory space.
2. These actions were coordinated.

At the same time, adding symbols characteristic of different eras to the ternary connective reduces the variability of interpretations. It increases the accuracy of the individual's "tuning" to the intuitive perception of the image originally embedded in the symbol. From epoch to epoch, the state of human consciousness changed (Table 4 in Chapter 9). Accordingly, the symbolic systems used to transfer knowledge also changed, which was studied by (Kovalyov et al., 2020). One of the symbols that accompanied man from the Paleolithic to the present is the World Tree (or the Tree of Life). Its images, characteristic for different epochs and the corresponding levels of channel activity, are known, and it is advisable to use them to build a ternary connective according to the scheme, according to Figure 6 of Chapter 3.

So, in Figure 1, a tree tattoo (modern era) evokes a positive emotional reaction and attracts attention; literary images and plots (New time) connect the mind and generate verbal reasoning about the intent and meaning, cabalistic, natural-philosophical, magical (Middle Ages, Antiquity, Ancient World) symbols activate the will and mind, initiating actions and evaluating reactions, mythical graphemes (pra-civilizations, Neolithic, Mesolithic) put the human ego in front of the World Tree in all its aspects, and the Paleolithic drawing relieves the already tuned consciousness of unnecessary details, gradually merging it with the Tree of Life as such.

Figure 1. An example of the correct construction of a ternary connective - human - symbols - Tree of Life. The numbers indicate the activated levels

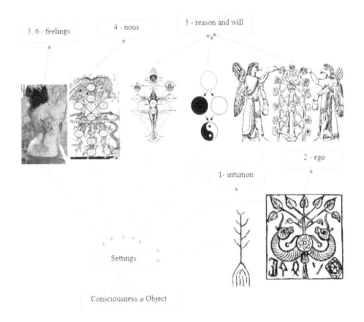

Of course, many of the symbols affect all or most of the channels; it is also quite possible that these influences are not coordinated, which reduces the accuracy of the "tuning" or even makes intuitive perception impossible. For example, Figure 2 indicates at least 4 plots - both mythical and fabulous, and it is very problematic here to tune in to the intuitive perception of the Tree of Life.

Knowledge of the 1st and 2nd kinds. Knowledge of the 1st kind, according to the definition given in Chapter 3, is the product of rational knowledge. Figures 1 and 2 can be subjected to historical-philosophical, cultural, and other research, as a result of which it is possible to accurately or presumably establish the dating and author, motives, meaning, the influence of the cultural environment, the public resonance of the work, and the like. It will be knowledge of the 1st kind.

There are also personal settings for intuitive perception, that is, some psychological "pointers" on the way to the individual experience of intuitive knowledge. This is knowledge of the 2nd kind. In this sense, the mutually agreed symbols in Figure 1 form such indicators, but the disagreed images in Figure 2 do not.

Figure 2. The plots "Tree of Life", "Grey Wolf", "Deer - Golden Horns", and "World Serpent" and the influences are not coordinated - intuitive perception is impossible (Sitnikova, 2020)

Let us give an example of using invariants, calibrations, and ternary connectives for the exegesis of the myth of the Deluge in its biblical version.

The plot of the myth is well known. Noah warned of the impending destruction of humanity mired in sins, built and populated the ark by the instructions given to him by God, crossed the waters of the Deluge, and landed on Mount Ararat, eventually giving rise to a new humanity. The invariants we are interested in are death, crossing waters as boundaries between worlds with different conditions, and the absence of perceptions when crossing.

Now let's recall the earlier Sumerian myth about Gilgamesh in its Akkadian version: Gilgamesh, wishing to gain immortality, goes to Uta-napishti, who survived the Deluge, caused by the gods to destroy sinful humanity, thanks to a ship built in advance and a warning from one of the gods, and the first ancestor of the new humanity. The invariants are the same.

It seems that the plot is very similar, but there are important details (calibrations): Gilgamesh is looking for Uta-napishti in the world of the dead, along the way he passes through dark fields, sees the light, and crosses the sea. Crossing a river or sea, as well as passing through forests or deserts, also with the help of a guide - a metapomp - (Slavic and Celtic mythology) - typical features of the "path" to the world of the dead (invariants), as well as a kind of "light at the end of the tunnel" (R. Moody).

So, the myth of Noah can be interpreted as the real death of Noah and the transfer of his soul through the boundary, "formless" and dangerous space to a new place of existence with different conditions. Let us accept this as a hypothesis and postpone the analysis of this most interesting myth until Chapter 10, where it will be compared with the expulsion of Adam.

Let us show a ternary connective constructed since the above reasoning; myth texts are replaced by pictures for clarity (Figure 3).

Figure 3. An example of a ternary connective: man - myth - posthumous scenario

RATING ASSESSMENT OF SOURCES

The Criteria Selection

7 criteria are determined by the proximity of the data to intuitive perception, the possibility of constructing a ternary connective, and the reliability of determining invariants and calibrations.

1. Antiquity. As the finds in the caves of La Ferracy (France) and Shanidar (Iran) showed, funeral rites, and, consequently, some ideas about the afterlife, took place even among the Neanderthals[1]. According to excavations at the site of Sungir (Russia) (Bader 1967) and Brno, in the caves of Grimaldi (Italy), and others, that already belong to modern humans, the existence of such a rite is already beyond doubt. These burials date back to the Upper Paleolithic, but earlier finds exist. Ideas about the afterlife arose at about the same time as the beginning of symbolic activity - when a person's intuition began to weaken and some help was needed to activate the intuitive channel (Kovalyov et al., 2020). By the time of the Upper Paleolithic, images like illustrations of ideas about the afterlife appeared, for example, Figures 4, and 5 allow for the following interpretation: the soul of a person killed by a wounded bison, exits through the head and flies away like a bird - but perhaps this reconstruction corresponds to later ideas.

Figure 4. Bird-man (a man with a bird's head) (n / d)

Figure 5. Bird-man). The right part of Figure 4(zoomed image) (n / d)

Such images correspond to a much higher activity of intuition compared to the modern period; on their basis, it is easy to construct a ternary connective; the brevity of the image facilitates the selection of invariants. However, due to the small number of findings, one cannot be sure of the correctness of the interpretation, and the possibility of using ethnographic analogies for verification is doubtful (Leroi-Gourhan, 2015).

Therefore, images accompanied by comments would be ideal, allowing to exclude or reduce the variability of interpretations - which already shifts the border of antiquity to the time of the emergence of writing, that is, at the time of the Eneolithic pra-civilizations and civilizations of the Ancient World, and, accordingly, pushes back from the time of the activity of intuition.

2. Tradition is always preferable to any single artifact. If rituals, images, and texts are created and practiced for a sufficiently long time, then the search for invariants and building ternary connectives on their basis becomes more reliable. However, there are dangers here: as the dominant channels change, the symbols cease to "work" - the "spirit" is replaced by the "letter", misunderstanding, heresies, and parodies arise. How, for example, can a modern human understand the following text: *"Introduction of the deceased to the gods. ...VII. I am the Phoenix in Heliopolis. I am the keeper of the scroll of the book of things that have been created and of things that are yet to be created. Who is this? This is Osiris. Others, however, say that it is the dead body of Osiris, and still others say that it is the stool of Osiris. The things that have been created and the things that will be created are related to the dead body of Osiris. Still, others say that things that have been created are Eternity, Things that will be created are Permanence, and that Eternity is Day and Permanence is Night".* (Ancient Egyptian Book of the Dead, 2003). The transla-

tor and Egyptologists are reenactors, the tradition is long dead. It is impossible to get a competent answer to the question "who is it". Therefore, the liveliness of tradition is of the utmost importance. The "Tibetan Book of the Dead" is still read over the dead; Buddhist priests (lamas) may be invited for consultation, which is what translators do. In this sense, the "Tibetan Book of the Dead" is preferable to the Ancient Egyptian.

3. Content. Several categories of sources can be distinguished here: directly devoted to the topic of the afterlife (Egyptian and Tibetan books of the dead), episodically concerning certain circumstances and events in the world of the dead (Inanna's Descent to the Netherworld, the Epic of Gilgamesh, the Bible, Popol Vuh), allegorically mentioning about such circumstances and events (the myth of the Flood and the like, as well as many fairy tales), rites and rituals, drawings (icon painting, symbols, rock art).

4. The definability of invariants and calibrations is an obvious criterion, especially important for sources that use allegories, symbols, rituals, and illustrations.

5. Type: text, image-symbol, ritual, cult objects, myth, fairy tale, as well as interpretations of modern authors for each of the types.

6. The role in the ternary connective is determined by the level of channels through which the artifact affects consciousness (there may be several of them), the accuracy of consciousness tuning (data certainty or blurring), belonging to invariants or calibrations.

7. Consistency with other sources. This is a rather subtle criterion: sources can talk about the same thing, using different symbols and focusing on different channels - they should be recognized as consistent, despite all the differences in form. A source may report some unique detail that others do not - here one should consider whether the information contained in it can be used as calibrations; if not, it should be ignored. A source can give information that is partially consistent with others, but simultaneously be unverifiable - it would be more careful to exclude it (like the data of Flammarion or Swedenborg). Finally, the source may contain an unconventional interpretation of the traditional doctrine, which is classified as heretical and probably based on some form of sophistic substitution of concepts (Aristotle, 1978).

A source's rating is necessary to determine whether it can be used in this study; here we restrict ourselves to a formal assessment, without setting out the content and without reconstructing the meaning of the sources (Chapters 13, 14).

Let's apply the following rating calculation (simplified compared to the traditional expert evaluation procedure): for each of the criteria, we will evaluate on a five-point scale (5 - fully meets the criterion; 4 - mostly meets; 3 - meets, but is unclear; 2 - meets, but is doubts about the reliability, authenticity or quality of the translation; 1 - unreliable, non-authoritative, non-interpretable); considering the criteria to be equivalent, the marks obtained must sum up.

Sources with a low rating are more careful not to consider this. Sources with a single rating for at least one criterion should also be excluded.

The Sources Evaluation

Upper Paleolithic artifacts (Abramova, 1971; Blednova et al., 1998; Leroi-Gourhan, 1967; 1982; 1992; 2015; Levi-Strauss, 1966; Lewis-William, 2002; Stolyar, 1985; Taylor, 2010/1871; Oliver, 2013; Zubov, 2017).

Characteristics for each of the 7 criteria:

1. The most ancient and closest to intuitive knowledge artifacts. Mark - 5.
2. Traces of the most ancient ideas about the afterlife are present both in modern religions and in cults (shamanism, the cult of ancestors, animism, fetishism) of locals in remote areas of the planet. Mark - 2.
3. Content - about posthumous existence and rebirth - probably and should be reconstructed; whether artifacts express posthumous relationships and events is not always clear. Mark - 2.
4. Taken in their entirety, they allow a simple definition of invariants and the construction of a ternary connective. Some artifacts fall out of this range and are interpreted hypothetically. Mark - 3.
5. Images-symbols (cave and rock art, ornaments on various items); cult objects (?) - items made of skulls, churingas, howler monkeys, "Paleolithic Venuses" - the connection between funeral rituals and the cult of fertility is considered probable); traces of funeral rites. There are no authentic explanatory texts. Mark - 3.
6. Impact mainly on channels 1, and 2 levels. Inaccuracy of tuning due to serious differences from the state of activity of channels in a modern person. Mark - 3.
7. Consistency of the basic idea of posthumous existence and subsequent rebirth in modern religions and cults; indefinite preservation (perhaps vulgarization, simplification, modification) in "folk" beliefs and customs. Mark - 3.

The total score is 21 points out of 35.

Mesolithic and Neolithic artifacts (Collins, 2014; Golan, 1993; Danylenko, 1986; Lhote, 1962; Mykhailov, 2005; Schmidt, 2020).

Characteristic by criteria:

1. Artifacts close in time to the 2nd and 3rd levels channels activity. Mesolithic artifacts are few, while Neolithic artifacts, on the contrary, are widespread. Mark - 5.
2. Neolithic ideas about posthumous existence are present in modern religions and cults of different peoples. Mark - 3.
3. The content - about the danger and falsity of the posthumous world, the need for the right choice, and struggle - probably should be reconstructed. The relation of artifacts to posthumous representations is more definite than in the previous case. Mark - 3.
4. Artifacts make it possible to determine invariants and construct a ternary connective. Calibrations are more difficult to determine. Mark - 3.
5. Images-symbols (rock and cave paintings, ornaments on various products), burial structures and complexes (menhirs, dolmens, cromlechs), objects of worship, as well as settlements with domestic and public sanctuaries, such as Göbekli Tepe and Chatal Huyuk). There are no explanatory texts. Mark - 4.
6. The impact is mainly on channels of 2-3 levels; the inaccuracy of tuning is due to serious differences from the state of activity of channels in a modern person. Mark - 3.
7. Satisfactory consistency in the idea of danger, struggle, and choice in modern religions and cults, possible vulgarization of calibrations in "folk" beliefs and customs. Mark - 3.

The total score is 24 points out of 35.

Artifacts of the Eneolithic and Pra-civilizations (before the reliable emergence of writing - the signs and symbols found in some cases can be interpreted as written monuments, but there are no generally recognized decipherments). (Watkins, 2005; Albedil, 1991; Mykhailov, 2005; Dergachev & Manzura, 1991; Korvin-Piotrovsky, 2008).

Characteristic by criteria:

1. Artifacts of the period of the 1st level channel activity decline and the 2-3 levels channels high activity (for the 3rd - close to the maximum). The heyday of complex symbolism, witchcraft practices, the period of creation of myths and fairy tales, primary religious cults, and hierarchies of numerous gods and demons. The possible creation of writing is the transition from pictograms to hieroglyphs, and, possibly, to signs (Vinci culture). Mark - 5.
2. Many achievements and ideas about the afterlife have entered modern culture. Mark - 3.

3. Sufficiently complex symbolism does not always give an unambiguous interpretation; possible written comments do not have generally recognized decipherments. This implies the need to reconstruct the content. The structure of the "three worlds" is probably clarified, and the possibility (including bodily) of movement between them, its dangers, and means of protection are reported. The similarity of some ideas with modern ones gives rise to the temptation to interpret from modern positions. Mark - 3.
4. Artifacts make it possible to reconstruct invariants; the reconstruction of calibrations is much more uncertain. Mark - 3.
5. Images-symbols, seals, ceramics, objects of worship, sanctuaries, burials, traces of funeral rites. The texts are not deciphered. Mark - 3.
6. A complex (magic) effect on the channels of the first three levels. Inaccuracy of tuning due to differences in their activity from the activity of the channels of a modern person. Mark - 4.
7. The preservation of the content in modern religions and cults, "folk" tales and customs, and literary versions of myths recorded in subsequent times, should be assessed as good. Let us again point out the danger of accepting the results of subsequent development or vulgarization of the original ideas. Mark - 4.

The total score is 25 points out of 35.

The Egyptian Book of the Dead, Its Antecedents and Successors (Shaposhnikov, 2003; Wallis Budge, 1967/1895).

From the time of the Old Kingdom's first dynasties (about 2625 BC), a written canon of the funeral service was recorded in 2355-2155 BC on the walls of the interior of the pyramids. A collection of such kingdom texts has been called the "Pyramid Texts" (Mercer, 2020/1952). The interregnum period (2150-2040 BC) gave rise to a new canon of mortuary texts for the nobility and rulers of the nomes - the "Coffin Texts ". These texts contain parts of the ancient "Pyramid Texts", but mostly consist of the writings of the priests of that time. Magical texts that opened otherworldly abodes to the deceased and told how to gain immortality began to be written down during the Middle Kingdom (circa 2010-1785 BC). Over time, the texts started to be placed inside the coffin. Depending on the status and wealth of the deceased, the texts were lengthy and short, with or without drawings, executed in calligraphy and carelessly (Figure 6).

Figure 6. The judgment of the dead in the presence of Osiris. Papyrus of Hunefer (n / d)

Mortuary texts of the New Kingdom, including parts of the "Pyramid Texts", " Coffin Texts" and texts of the priests of the New Kingdom (1550-1070 BC) received the name "Book of the Dead"; its last editions belong to the Sais dynasty (663-525 BC). In addition, there were texts similar in content "The Amduat" (That Which is in the Afterworld), "The Book of Gates", etc. Because mortuary texts existed in oral transmission before they were fixed on the walls of the pyramids, the Book of the Dead reflects the development of the mortuary service by the priests of Abydos, Panopolis, Hermopolis, Herakleopolis, Memphis, Heliopolis, Busiris and Buto, but not Thebes, at least in for 2500 years.

This implies the absence of a canonical edition of the book and its very conditional structuring. Usually, 4 parts are distinguished: 1st: chapters from 1 to 16, accompanying the procession of the funeral procession to the cemetery, prayers for the "exit (of the deceased) in the afternoon" and hymns to the gods Ra and Osiris; 2nd: chapters 17-63 - a description of the rites of the rebirth of the "exit (of the deceased) during the day", victory over the forces of darkness, weakening the enemies of the deceased, gaining power over the elements; 3rd: 64-129 chapters - a description of the rites that accompanied the transformation of the deceased into a deity, his introduction to the Boat of Millions of Years, knowledge of various mysteries, return to the tomb and the Afterlife Court; 4th: chapters 130-162 - a description of the magical rites that ensure the safety of the mummy and funeral texts glorifying the name of the deceased. These texts were read during the year after death on certain holidays and on the days of offering gifts to the deceased (Ancient Egyptian Book of the Dead, 2003).

Characteristic by criteria:

1. Texts and illustrations dating back to the period of levels 2-3 channels high activity, the increase of the 4th level activity, and the weakening of the activity. Religion, complex hierarchies of polymorphic relative gods opposing intuition monsters, magic, and a funeral cult permeate all aspects of society as ever. The creation and evolution of writing - the transition from pictograms to hieroglyphs and signs. The construction of the ternary connective does not lack material. Mark - 5.
2. Many achievements and ideas about the afterlife have entered the modern culture, especially in the ideology and rituals of the Masonic lodges; indirectly, the tradition is alive. However, the original rituals are not performed, and speeches are not delivered. Mark - 4.
3. Completely devoted to the wanderings of the soul in the afterlife. The complex symbolism of sayings, images, objects, and rituals is accompanied by detailed written comments, often giving different interpretations. One can trace the evolution of the content - the increasing role of magical influences from the world of the living on the posthumous fate. Mark - 5.
4. Basic ideas - invariants (posthumous existence, judgment by the gods of life events, full of dangers until the final death of the path to the light, the need for help to overcome it) - very persistent, differences in comments (calibration), although significant, do not affect the basic ideas. Both invariants and calibrations are relatively easy to determine. Mark - 4.
5. Type: texts, images, temples, tombs, mummies, collections of funerary things. Mark - 5.
6. A complex effect on channels of all levels, especially - 1-4. Inaccurate settings are due to the ambiguity of symbols, variability of comments, and differences in the activity of channels from a modern human. Mark - 4.
7. Good preservation of ideas about the influence of lifetime actions on the outcome of the court of the gods, as well as the patronage of the gods and the help of rituals in dangerous posthumous wanderings in modern religions and cults, literary versions, and direct finds. Mark - 4.

The total score is 31 points out of 35.
Sumerian-Akkadian Myths (Foster, 2001; Dedovich, 2020).
Sumerian legends about Gilgamesh were presumably formed at the end of the first half of the 3rd millennium BC. e., although the records that have come down to us are 800 years younger. Records of other myths also belong to the same time: About the Deluge, Inanna's Descent to the Netherworld, and others, as well as the first surviving records of an Akkadian poem about Gilgamesh from the 2nd mil-

lennium BC. Fragments of another version of this poem, as well as fragments of its Hurrian and Hittite translations, have been found in Palestine and Asia Minor. Finally, from the 7th and 6th centuries BC, the text of the poem's final "Nineveh" version has come down to us (Foster, 2001).

Characteristic by criteria:

1. Texts dating back to the levels 2-3 channels high activity period, the decline in the activity of intuition, and the strengthening of the 4th level channels. Gods, demigods, mystical and demonic beings living among people, beings who get alive to the world of the dead and talk about its, magic (Fosse, 2016). However, the mortuary cult never reached the scale characteristic of ancient Egypt and is known only in fragments. The creation and evolution of writing - the transition from pictograms to cuneiform signs. Constructing a ternary connective is difficult, but unique details (calibrations) that are important for interpreting other data have been preserved. Mark - 4.
2. Some ideas about the afterlife have entered the modern culture, rather folk (myths, fairy tales) than official religious, and magical practices up to modern ones. At the same time, we note that in the list of skills of Mesopotamian priests, the spell of the dead or rituals that affect the afterlife do not appear (Fosse, 2016). Mark - 3.
3. The texts themselves are clear enough; they give the concept of deceit as a property of the world of the dead and its inhabitants; it is difficult for a living being to understand the meaning of their speeches. It is possible to get into the world of the dead alive - either by the gods' decision or with the help of magical amulets. Getting out of there alive is even more difficult - you need supernatural help or redemption of your soul with its replacement with the soul of another creature, not necessarily of the same rank. It is also important to note that Inanna is the goddess and sister of the mistress of the underworld, and Gilgamesh is "two-thirds god and one-third man", so their relationship with the world of the dead can be very different from the relationship with this world of mere mortals. Mark - 3.
4. The invariants are very archaic, most likely, they still have a Neolithic origin, and they are already out of the attention of the authors, who focus on various details that are interesting for the ternary connective. Rating - 3.
5. Texts, images, cult objects, tombs, and funerary items. Mark - 4.
6. Data fragmentation - especially in comparison with Ancient Egypt. Impact on all channels, especially levels 1-4. Differences from the consciousness of modern man. All this complicates the construction of a ternary connective. Mark - 3.
7. Preservation in texts, modern magical practices, some myths of later religions, folk myths, and fairy tales. Mark - 3.

The total score is 23 points out of 35.

Monotheistic religions. Historically, polytheistic religions preceded monotheism (Eliade, 1981-1988; Tokarev, 2005). Accordingly, their ideas about posthumous existence largely retained the features of previous periods; their features are well illustrated by the Egyptian Book of the Dead and the Sumerian-Akkadian myths. There is no need to supplement them with new materials.

Pharaoh Akhenaten is believed to have first attempted to introduce a monotheistic cult but was unsuccessful. In the future, the image of the one God will become the basis of the Abrahamic religions - Jewish, Christian, and Islamic- and will be interesting in developing religious experience compared to polytheism. Despite the common basis, Judaism, Christianity, and Islam are significantly different from each other in understanding God and His relationship with a human but are close to the events of the posthumous existence of the soul in the sense that the emphasis is on the Last Judgment and its final state.

We will use the electronic version (Holy Bible. ASV).

The tradition is both ancient and alive, dating back at least 3500 years, actively fighting the polytheistic influences of the beliefs of Egypt, Mesopotamia, Palestine, the Hellenistic ecumene, as well as ancient and modern atheism, for the adepts - based on Divine revelation.

Characteristic by criteria:

1. Texts dating back to the period of high activity of channels of levels 2-3 continue to be created during the deactivation of levels 1-3 channels and the growing role of channels of level 4. They contain direct statements on death and the afterlife and hidden ones that need interpretation. Mark - 4.
2. All ideas about the afterlife have firmly entered modern official and popular culture. Mark - 5.
3. Some texts are quite clear, and the comments of the bearers of the tradition help to understand others - but these comments differ significantly (calibrations), despite the common origins - ideas about the creation of man by God, death as a result of the fall, the Savior, reunification with the body (new body), Terrible judgment at the end of time, the outcome of which depends both on the actions of a person and on the grace of God, heaven or hell as final states, the preservation of the individual in hell and heaven (invariants). Mark - 4.
4. The rack and calibration invariants are very different, which makes it difficult to construct a ternary connective. Mark - 4.
5. Texts, icons, objects of worship, tombs, and funeral rites - differ for different faiths. Mark - 4.

6. Consistency of invariants facilitates the construction of a ternary connective and a complex effect on all channels. Differences from the consciousness of a modern person, which is visible when comparing the original source with relatively modern comments, as well as the inconsistency of calibrations, create certain difficulties. Mark - 4.
7. Texts, modern religious practices, monastic experience, and folk customs are some of the foundations of contemporary culture. Mark - 5.

The total score is 30 points out of 35.

Popol Vuh (Popol Vuh, 2007) is interesting not only as an example of the tradition of the Quiche people but also as a description of the mores of the rulers of the underworld, as well as a detailed account of the successful struggle against them by the divine twins Hunahpu and Xbalanque. Therefore, despite the late record (already after the Spanish conquest) and the noticeable influence of Christianity, it is advisable to include the text in the list of evaluated sources.

Characteristic by criteria:

1. Representations that arose during the Stone Age (in principle, lasted in Mesoamerica until the very Spanish conquest), preserved (how authentically?) in the memory of the author (or translator into Spanish, if the original existed), covering periods of activity and decline in the activity of channels 1 -3 levels recorded during the dominance of the 4th level channels. Mark - 2.
2. Ideas about the afterlife are localized in folk customs and the descendants of several Indian peoples of Mesoamerica - do not forget that this is a book of the Quiche people, who are in hostile relations with neighbors who could have other ideas that have not come down to us. Mark - 2.
3. The text is clear enough from the point of view of the data interesting for this study (invariants and calibrations). Contains a large amount of data about the afterlife and the struggle with its masters. Mark - 3.
4. Invariants are determined by comparison with other materials, the calibrations are original. Mark - 3.
5. Text, images, religious objects, tombs. Mark - 3.
6. The text links several legends, which helps build a ternary copula. The differences from the consciousness of a modern person are quite large, despite the Christian upbringing of the author, who may have modified authentic traditions, which is a complicated circumstance. Mark - 1.
7. Preservation of the text - only the Spanish version, as well as folk customs little known outside of Mesoamerica. Mark - 2.

The total score is 16 points out of 35.

Myths and fairy tales. Modern researchers have done much work on their fixation and interpretation (Kun, 1922; Tokarev, 1987; Shirokova, 2004; Propp, 1986; Abrahamsson, 2009/1951; Paulme, 1967; Mwania, 2016; Baloyi & Makobe-Rabothata, 2018; Berndt & Berndt,1988/1964). Separately, we single out the myths related to the cult of ancestors, which gradually evolved from domestic and tribal traditions to the basis of official cults in Japan and China (Ermakova, 2002; Keightley; Zakurdaev, Georgievsky, 2022). It is precisely this already interpreted information about posthumous existence that it is advisable to use: the price of a misunderstanding of the original records is much higher than the price of possible inaccuracies of interpreters.

Characteristic by criteria:

1. Representations that arose, possibly, back in the Stone Age, were preserved (and distorted) in people's memory and continued to be created for a very long time, covering periods of activity and decline in the activity of channels 1-3 levels, recorded during the dominance of channels 4- go and even 5-6 levels. They contain a certain number of direct, but more hidden and need interpretation data about death and afterlife existence. Mark - 2.
2. They are, to a greater or lesser extent, folk, unconscious ideas about the afterlife. Mark - 3.
3. All texts need interpretation to reconstruct important details (calibrations). The invariants are not original. Mark - 2.
4. Only to clarify some details (calibrations). Mark - 2.
5. Recorded in the recent past or recent decades, stories of people who do not always understand the meaning of the myths and fairy tales known to them, were created much earlier. Mark - 2.
6. Fragmentation and inconsistency of data. The possibility of their use only after comparison with other materials. Mark - 2.
7. They are consistent with (explain) some folk beliefs and customs, contain (not always) monotheistic paraphernalia, and contain some (modified and vulgarized) representations of the periods when they were created. Thanks to the wide dissemination of myths and fairy tales, many details of these representations have come down to us. Mark - 2.

The total score is 13 points out of 35.

The Tibetan Book of the Dead directly describes the states and events of the afterlife. It is a Buddhist text read over the deceased's body, bearing traces of the influence of the Bon religion, a pre-Buddhist belief of the Tibetans (Thurman, 2011; Tibetan Book of the Dead, 1992). There are comments on contemporary authors (Jung, 2002) and tradition bearers (Thurman, 2011).

Characteristic by criteria:

1. Representations that arose in very remote times (presumably covering periods of activity and decline in the activity of channels of levels 1-3), subjected to Buddhist interpretation and recorded relatively late, presumably, during the period of dominance of channels of the 4th level. Mark - 5.
2. Due to the late record and the living tradition, it has a clear structure and contains a consistent and even rational description, quite clear both for the bearer of the tradition and the person who has an idea about it. Mark - 5.
3. Directly devoted to the posthumous theme. The text contains a detailed description of the events and states of the afterlife, expressed in symbolic form, but quite understandable; some editions are provided with illustrations that reproduce Tibetan originals, the symbolism of which is comprehensible to adherents of Buddhism. Mark - 5.
4. The invariants are determined from the text itself and modern commentaries. Calibrations outside the context of the Tibetan version of Buddhism are far from always clear. Mark - 4.
5. Texts, images, and objects of cult, historically and still used in the posthumous ritual. Mark - 5.
6. It affects all channels, from intuition to sensual sensations, generating sensations from enlightenment to strong emotions, which greatly facilitates the construction of a ternary bundle. Rating - 5.
7. The consistency of invariants and calibrations in the distribution areas of Buddhism and Hinduism can be assessed as high and average compared to other materials. Mark - 4.

The total score is 35 points out of 33.

Let's summarize all the ratings in the final Table 1 and determine the ranks of the sources.

Table 1. Ranking of sources by posthumous existence

Rank	Rating	Source
1	33	Tibetan Book of the Dead
2	31	Egyptian Book of the Dead
3	30	Monotheistic religions
4	25	Artifacts of the Eneolithic and pra-civilizations
5	24	Artifacts of the Mesolithic and Neolithic

continued on following page

Table 1. Continued

Rank	Rating	Source
6	23	Sumerian-Akkadian myths
7	21	Artifacts of the Upper Paleolithic
8	16	Popol Vuh
9	13	Myths and fairy tales

Thus, in general, and when determining invariants, one can trust sources with higher ranks more. When analyzing details, one can be guided by a rating according to the necessary criterion, not forgetting that any methods of physical measurements cannot verify all these sources.

FUTURE RESEARCH DIRECTIONS

Immediate tasks: Using the ranking sources, it is necessary to reconstruct the post-mortem scenarios and compare them with each other, as well as the generalized post-mortem scenario substantiated in Chapter 11. These tasks will be solved in Chapters 13-14.

CONCLUSION

Let us summarize the results obtained in three theses and one reasoning.

Thesis one: pragmatic. From the viewpoint of this work, the tasks were solved: a method based on the use of the apparatus of ternary connectives was created, which allows the interpreting of symbols, texts, plots, signs of different eras, ranking sources (Table 1) and reconstructing the ideas of different eras about the structure of a posthumous creature. According to the sources, these results will be used in Chapters 13-14 to reconstruct post-mortem structures and events and to compare with the general post-mortem scenario substantiated in Chapter 11.

Thesis two: proofing. For followers of intuitionism, verification using intuitive knowledge obtained with the help of a ternary connective is convincing. For humanitarian, finding similar structures and scenarios would be believable because texts and artifacts are processed using ternary connectives (K1). Similarly, synchronized psychological changes (K2) would be convincing for psychologists. For representatives of the natural sciences, experimental proof is necessary.

Thesis three: general. The created method has a paradigmatic significance: hermeneutical interpretations are faced with discrepancies between modern thinking, conceptual and linguistic apparatus, and methods with ancient correlates. This discrepancy is so great that modern researchers must use even occult terminology, which is unacceptable. The presented method allows both to eliminate this inconsistency and to use precisely defined terminology.

Let's preliminary test the developed method, choosing data on communication with the "world of the dead" for this purpose. We use the data of Flammarion (Flammarion, 1920-1922), a known episode in the Odyssey (these sources were not ranked at all), as well as Table XII of the Epic of Gilgamesh and the story about the Witch of Endor in the Bible (these sources are ranked).

Cases of such communication have always been regarded as mystical; nevertheless, the sources contain informative descriptions. Let's group this data and try to use it for preliminary testing of a general post-mortem scenario and for determining the role of the ternary connective.

1. Generally, communication is not possible. Very rarely, however, it occurs - usually with those who died a few hours or days ago; as the date of death moves away, the frequency of communication decreases. Close relatives often communicate with the dead, and less usually old friends (Flammarion, 1920-1922). Both the deceased and his addressees are interested in such communication.
2. Anyone who wants to communicate with the long dead has to resort to certain actions: to have a strong need, to come to the right place at the right time, to concentrate, to make a sacrifice, to cast a spell, etc. All these necromantic actions are rated as extremely dangerous.
3. The information received is of two types. When communicating with the recently deceased, it is specific (for example, about the place where the deceased hid the will, about the murderers, about the place and circumstances of death, and so on). When communicating with the long dead, information refers to events of the distant past or indefinite future – usually, it is expressed very vaguely, symbolically, without reference to a specific time and place; however, there are exceptions.
So:

Point 1. The impossibility of communication is explained by the decrease in the potential of all channels of subjective space in the dead, so much so that it becomes less than the threshold of sensitivity in the living. An aging person experiences problems with vision, hearing, etc. - this process continues here. Moreover, as the convolution progresses, the subjective space loses the channels through which communication is possible - first sensual, then rational (Table 4, Chapter 11).

Nevertheless, the emerging communication can be explained by the emergence of "resonance", which is facilitated by both the mutual desires of the living and the dead and the similarity of the mental organization of close relatives and friends. Love, like intuition, is found outside of time...

Point 2. We do not undertake to judge the effectiveness of the listed necromantic actions, but their goals are quite reasonable: creating conditions for "resonance" and increasing the potential for the dead. The danger is also understandable: the subjective space of the deceased, having increased its potential, but disrupted during convolution, acquires the ability to influence the subjective space of the living - and this impact is only destructive.

Point 3. Each type of data received should be evaluated in its own way. As long as the ideas about space and time and the channels of the 4th level of subjective space are not destroyed, the deceased can still transmit "rationally" information defined in space and time. The long-dead have lost these channels and concepts due to convolution and can only transmit intuitive knowledge through a ternary connective. Hence their symbolism and uncertainty.

So, the method has been successfully tested: it can be used to give plausible explanations for communication with the "world of the dead". It can also be used to study similar texts. There is no need for any mysticism here. However, even the formal correctness of the method does not yet mean the truth of the result - everything depends on the source's reliability. But how precarious are the sources! We need to remember this in the following chapters.

We note one more circumstance. The degree of reliability increases if the text or artifact meets a critical need - then it is carefully checked. The next sign is society's readiness to invest resources in a project related to a text or artifact. So, the ancient Egyptians invested resources in constructing tombs and pyramids, and not, for example, in purchasing weapons or agricultural implements. Finally, it is the duration of the concept's existence, which is impossible without tangible results. But it needs to change over time, and stopping the construction of pyramids or tombs does not mean recognizing the wrong ideas. So, in our post-industrial times, the presence of abandoned roads and factories does not mean their uselessness or the fallacy of projects - just going into virtuality becomes more effective by the ratio of pleasures and costs.

From this point of view, all the sources considered are reliable for ancient people; for modern people, they are a priori doubtful.

REFERENCES

Abrahamsson, H. (2009). *The origin of death: studies in African mythology*. Cambridge University Press. (Original work published 1951)

Baloyi, L., & Makobe-Rabothata, M. (2014). The African conception of death: A cultural implication. *Toward Sustainable Development through Nurturing Diversity*. https://doi.org/DOI: 10.4087/FRDW2511

Berndt, R. M., & Berndt, C. H. (1988). *Catherine Helen (Author) The World of the First Australians: Aboriginal Traditional Life: Past and Present*. Aboriginal Studies Pr. (Original work published 1964)

Bird-man (a man with a bird's head). [Illustration]. Cave of Lascaux. France. (n.d.). Yandex.com. Retrieved August 14, 2024, from https://yandex.com/collections/card/5a9c60b22321f21565f700e4/

Christenson, A. J. (2007). *Popol Vuh: The Sacred Book of the Maya*. University of Oklahoma Press.

Coleman, G. (Ed.). Jinpa, T. (Ed.), Dorje, G. (Transl.), Dalai Lama (Comment.). (2007). *The Tibetan Book of the Dead: First Complete Translation*. Penguin Classics

Collins, A. (2014). *Göbekli Tepe: The Origin of the Gods*. Bear & Company.

Dedović, B. (2020). *"Inanna's Descent to the Netherworld": A centennial survey of scholarship, artifacts, and translations*. Digital Repository at the University of Maryland. https://doi.org/DOI: 10.13016/ur74-yqly

Charon. Illustration of Dante's Divina Commedia Author: Doré, R. G. [Illustration]. *File: Charon by Dore.Jpg*. (n.d.). Wikimedia.org. Retrieved August 14, 2024, from https://commons.wikimedia.org/wiki/File:Charon_by_Dore.jpg

Eliade, M. A (1981-1988). *History of Religious Ideas*. (3 vol.). University of Chicago Press.

Foster, B. R. (2001). *The Epic of Gilgamesh*. W.W. Norton & Company.

Gadamer, H.-G. (1989). *Truth and Method* (2nd ed.). Continuum Intl Pub Group.

Goelet, O. (Transl.), Faulkner, R. (Transl.), Andrews, C. (Pref.), Gunther, J. D. (Intro.), & Wasserman, J. (Foreword). (2015). *Egyptian Book of the Dead: The Book of Going Forth by Day: The Complete Papyrus of Ani*. Chronicle Books. Flammarion, C. (1920-1922). *La mort et son mystère*. Editeur Fantaisium.

Hero mastering a lion. Relief. Palace of Sargon II at Khorsabad, 713–706 BC. Collection: Louvre Museum. Department of Oriental Antiquities, Richelieu, ground floor, room 4. Accession number: AO 19862. Photographer: Jastrow (2006). [Illustration]. *File: Hero lion Dur-Sharrukin Louvre AO19862.jpg.* (n.d.). Wikimedia.org. Retrieved August 14, 2024, from https://commons.wikimedia.org/wiki/File:Hero_lion_Dur-Sharrukin_Louvre_AO19862.jpg

Holy Bible - American Standard Version. (2019, April 1). Holy Bible - American Standard Version - ASV. https://holy-bible.online/asv.php? Landscape with Noah, Offering a Sacrifice of Gratitude. J.A. Koch, K.G. Chic (1803). Shtedel Art Institute. 2nd floor, Kunst der Moderne. ID 767 Source: The Yorck Project (2002). [Illustration]. *File: Joseph Anton Koch 006.jpg.* (n.d.). Wikimedia.org. Retrieved August 14, 2024, from https://commons.wikimedia.org/wiki/File:Joseph_Anton_Koch_006.jpg?uselang=ru

Joshi, K. L. (Sanskrit Text, English Transl.). (2005). *112 Upanishads.* (2 vol.) Parimal Publications.

Jung, C.-G. (2002). Psychological Commentary on The Tibetan Book of the Dead. In *Jung on Death and Immortality*. Princeton University Press., DOI: 10.1515/9780691215990-004

Keightley, D. N. (2004). *The Making of the Ancestors: Late Shang Religion and Its Legacy.* In *Religion and Chinese Society* (Lagerwey, J. Edit.). (2 vol.). The Chinese UP DOI: 10.2307/j.ctv1z7kkfn.4

Kovalyov, Y., Mkhitaryan, N., & Nitsyn, A. (2020). *Self-organization of the human mind and the transition from paleolithic to behavioral modernity.* IGI Global. DOI: 10.4018/978-1-7998-1706-2

Leroi-Gourhan, A. (1967). *Treasures of Prehistoric.* Harry N. Abrams.

Leroi-Gourhan, A. (1982). *The Dawn of European Art: An Introduction to Palaeolithic Cave Painting.* Cambridge University Press Leroy- Gourhan, A. (1992). *L'art pariétal - Langage de la préhistoire.* Jérôme Million

Leroi-Gourhan, A. (2015). *Les religions de la préhistoire* (7th ed.). PUF.

Levi-Strauss, C. (1966). *The Savage Mind.* University of Chicago Press.

Lewis-William, D. (2002). *The Mind in the Cave.* Thames and Hudson.

Lhote, H. (1959). *The search for the Tassili frescoes: the story of the pre-historic rock paintings of the Sahara.* Dutton.

Mercer, S. A. B. (2020). *The Pyramid Texts*. Global Grey. (Original work published 1952)

Mwania, P. (2016). Interface between African's Concept of Death and Afterlife and the Biblical Tradition and Christianity. (N.d.). Tangaza.Ac.Ke. Retrieved August 14, 2024, from https://repository.tangaza.ac.ke/server/api/core/bitstreams/2ad4ce8c-5d88-496a-8986-2b23b084f309/content

Noah's Ark. Islamic miniature. Before XVIII c. [Illustration]. *File: Noah islam 2.jpg*. (n.d.). Wikimedia.org. Retrieved August 14, 2024, from https://commons.wikimedia.org/wiki/File:Noah_islam_2.jpg?uselang=ru

Paulme, D. (1967). Two Themes on the Origin of Death in West Africa. *Man*, 2(1), 48–61. DOI: 10.2307/2798653

Pomeroy, E., Mirazón Lahr, M., Crivellaro, F., Farr, L., Reynolds, T., Hunt, C. O., & Barker, G. (2017). Newly discovered Neanderthal remains from Shanidar Cave, Iraqi Kurdistan, and their attribution to Shanidar 5. *Journal of Human Evolution*, 111, 102–118. DOI: 10.1016/j.jhevol.2017.07.001 PMID: 28874265

Riera, J. J. *Semiotic Theory*. Press Books.

Schmidt, K. (2020). *Sie bauten die ersten Tempel: Das rätselhafte Heiligtum am Göbekli Tepe*. C.H.Beck.

Taylor, E. B. (2010). *Primitive Culture. Researches into the Development of Mythology, Philosophy, Religion, Art, and Custom*. Cambridge University Press. (Original work published 1871), DOI: 10.1017/CBO9780511705960

The Deluge. R. G. Dore. [Illustration]. *File: World destroyed by water.Png*. (n.d.). Wikimedia.org. Retrieved August 14, 2024, from https://commons.wikimedia.org/wiki/File:World_Destroyed_by_Water.png

The judgment of the dead in the presence of Osiris. [Illustration]. Papyrus of Hunefer. Ancient Egypt. Collection. British Museum. Accession number EA 9901. Author: Unknown author. Wikipedia contributors. (n.d.). *File: The judgement of the dead in the presence of Osiris.jpg*. Wikipedia, The Free Encyclopedia. https://en.wikipedia.org/wiki/File:The_judgement_of_the_dead_in_the_presence_of_Osiris.jpg

Thurman, R. (2011). *The Tibetan Book of the Dead: Liberation Through Understanding in the Between*. Bantam.

Van Dijk, T. A. (2006). Discourse, context and cognition. *Discourse Studies*, 8(1), 159–177. Advance online publication. DOI: 10.1177/1461445606059565

Wallis Budge, E. A. (1967). *The Book of the Dead. The Papyrus of Ani*. Dover Publications. (Original work published 1895)

Watkins, T. (2005). *From Foragers to Complex Societies in Southwest Asia*. Kapitel 6. In *The Human Past: World Prehistory & the Development of Human Societies*. Thames & Hudson.

ADDITIONAL READING:

Abramova, Z. A. (1971). *Pervobytnoe iskusstvo*. [Primitive Art]. NSU. Albedil, M. F. (1991). *Zabytaja cyvylyzacyja v dolyne Ynda*. [A Forgotten Civilization in the Indus Valley]. Nauka. Aristotle. (1978). *O sofystycheskykh oproverzhenyjakh*. [On sophistical rebuttals] In *Aristotle. Works in four volumes*. v.2., 535-593. Thought.

Bader, O. N. (1967). *Pogrebeniya v verhnem paleolite i mogila na stoyanke Sungir*. [Burials in the Upper Paleolithic and a grave at the Sungir site]. *Soviet Archeology 1967*(3). https://arheologija.ru/bader-pogrebeniya-v-verhnem-paleolite-i-mogila-na-stoyanke-sungir/

Bakhtin, M. M. (1979). *Estetyka slovesnogho tvorchestva*. [Aesthetics of verbal creativity]. Art.

Blednova, N. S., Vishnyatsky, L. B., Goldshmidt, E. S., Dmitrieva, T. N., & Sher, Ya. A. (1998). *Pervobytnoe iskusstvo: problema proyskhozhdenyja* [Primitive art: the problem of origin]. KPI.

De Saussure, F. (1977). *Trudy po yazykoznaniyu* [Works on linguistics]. Progress.

Ermakova, L. M. (2002). Poklonenye predkam v japonskoj kuljture. [Ancestor worships in Japanese culture] In *Shinto - The Way of the Japanese Gods*. Hyperion.

Fosse, S. H. (2016). *Assyryjskaja maghyja* [Assyrian magic]. Eurasia.

Golan, A. (1993). *Myf i simvol*. [Myth and symbol]. Russlit. Danylenko V.M. (1986). *Kamyana Mohyla*. [Stone tomb]. Naukova Dumka. Dergachev, V. A., & Manzura I. V. (1991). *Poghrebaljnye kompleksy pozdnegho Trypoljja*. [Burial complexes of late Trypillia]. "Shtiintsa".

Korshunov, A. M. (1991). Kateghoryja poznavateljnogho obraza. [Category of cognitive image] In *Theory of Knowledge. Socio-cultural nature of knowledge*. Nauka.

Korvin-Piotrovsky, A. G. (2008). *Trypiljsjka kuljtura na terytoriji Ukrajiny* [Trypillia culture on the territory of Ukraine]. Institute of Archeology of the National Academy of Sciences of Ukraine.

Kuhn, N. (1922). *Leghendy y myfy Drevnej Ghrecyy*. Legends and myths of Ancient Greece. Public Domain. https://bookscloud.ru/books/80264

Losev, A. F. (1982). *Znak. Symvol. Myf* [Sign. Symbol. Myth]. Publishing House of Moscow University.

Losev, A. F. (1993). *Fylosofyja ymeny*. [Name philosophy]. In *Being. Name. Space. Thought*.

Lotman, Y. M. (2000). *Semiosfera* [Semiosphere]. Art-SPB.

Mykhailov, B. D. (2005). *Petroghlify Kam'janoji Moghyly: Semantyka. Khronologhija. Interpretacija* [Petroglyphs of Kamyana Mohyla: Semantics. Chronology. Interpretation]. MAUP.

Oliver, J. E. (2013). *Tajny jazycheskykh boghov. Ot bogha-medvedja do zolotoj boghyny.* [Secrets of the pagan gods. From the Bear God to the Golden Goddess]. Veche Fylosofskyj encyklopedycheskyj slovarj. (1983). [Philosophical encyclopedic dictionary]. Soviet Encyclopedia.

Popol-Vuh. (1993). Nauka. [Popol-Vuh]

Propp. V. Ya. (1986). *Ystorycheskye korny volshebnyh skazok.* [The historical roots of fairy tales]. Publishing House of Leningrad State University

Shaposhnikov, A. K. (Ed.). (2003). Drevneeghypetskaja Knygha Mertvykh. Slovo ustremlennogho k Svetu. The Ancient Egyptian Book of the Dead. The word of one striving towards the Light. Eksmo Publishing LLC

Shirokova, N. S. (2004). *Myfy keljtskykh narodov*. [Myths of the Celtic peoples. AST, Astrel, Tranzitkniga Sitnikova, S. A. (2020). *Drevnye sakraljnye znachenyja obraza khvojnogho dereva (ely, sosny) v tverskoj tradycyonnoj kuljture.* Ancient sacred meanings of the image of a coniferous tree (spruce, pine) in the Tver traditional culture. *Art education and science. 2* (23), 163-175 https://gardenmodern.ru/drevnie-sakralnye-smysly-obraza-hvojnogo-dereva-el-sosna-v-tverskoj-tradicionnoj-kulture/

Sovremennyj fylosofskyj slovarj. (2003). [The Newest Philosophical Dictionary]. Book House.

Stolyar, A. (1985). *Proyskhozhdenye yzobrazyteljnыkh yskusstv.* [The origin of the visual arts]. Art.

Tibetskaya kniga mertvyh. (1992). [Tibetan Book of the Dead] (Transl.). Chernyshev Publishing House.

Tokarev, S. A. (2005). *Relyghyja v ystoryy narodov myra* [Religion in the history of the peoples of the world]. Respublika.

Tokarev, S. A. (Ed.). (1987-1988). *Encyklopedyja «Myfy narodov myra»*. [Encyclopedia "Myths of the peoples of the world"]. (2 v.). Sovetskaya entsiklopediya.

Vasiliadis, N. (2012). *Taynstvo smerti*. [Sacrament of death]. Publishing House of the Holy Trinity Sergius Lavra. https://pravoslavnoe.uaprom.net/p782708550-vasiliadis-tainstvo-smerti.html

Chapter 4
Structures Checking

ABSTRACT

The method of post-mortem scenario verification (Chapter 3) was applied to compare the structures of the sensory space with the structures of the human personality, reconstructed following the ideas of different eras about post-mortem existence. Concepts of various eras are related and compared; the soliton-wave model is a "coordinate system". Their invariants are defined. This makes it possible to reconstruct post-mortem scenarios and compare them with each other and with the generalized post-mortem scenario substantiated in Chapter 2.

BACKGROUND

In the sources reflecting the general human experience, various terms express multiple structures of the human personality. To test post-mortem scenarios (Chapter 2), it is necessary to compare these terms with the structures of subjective space and each other to find out: what structures (levels, channels) correspond to? How do they relate to the same personality? How do they interact with the environment? How are they comparable to each other? And whether they are "static" or reflect the dynamics of post-mortem evolution.

For such comparisons, carried out according to the methodology developed in Chapter 3, a "coordinate system" is needed. The human-environment model will be used in this role. This allows for solving all the above issues in a complex.

The sources, ranked in Chapter 12, are related to different epochs.

First, the ancient times before the emergence of full-fledged writing (Leroi-Gourhan, 1967; 2015; Danylenko, 1986; Dergachev & Manzura, 1991; Levi-Strauss, 1994; Eliade, 2002; Lewis-William, 2002; Mykhailov, 2005; Korvin-Piotrovsky, 2008; Zubov, 2017).

DOI: 10.4018/979-8-3693-9364-2.ch004

Second, the pre-civilizations and ancient civilizations of Mesopotamia and Egypt (Wallis Budge, 11967/1895; Albedil, 1991; Foster, 2001; Dyakonov, 2006; Granin, 2011; Collins, 2014; Goelet et al., 2015). Sources have been characterized in Chapter 3, we additionally note that it should be remembered that the texts of the Sumerian myths have come down to our time in a very bad condition; in fact, each published translation is a reconstruction, "glued together" from many fragments (Dedović, 2020). Therefore, the translations differ, sometimes quite significantly. A similar remark can be made about the Egyptian Book of the Dead. In such a situation, the authors tried to use several translation options to obtain the most complete version. But at some point, the versions began to diverge too much, and here it was necessary to stop.

Third, the Abrahamic religions (Holy Bible (ASV); Philokalia, vols. 1-5, 1992; St. Palama, 2011; Vasiliadis, 2012; St. Luke (Voino-Yasenetsky), 2013; St. Theophan, n / d; Human Tripartism, n / d). Here the accuracy of the Hebrew Bible canonical text is much higher - thanks to the efforts of the Masoretes, the text was corrected in the early Middle Ages. However, Christian translations have been made from earlier versions (and translations) and they are already significantly different. In addition, there are confessional differences. Therefore, it is necessary to refer to various translations and commentaries here.

Fourth, the Tibetan Book of the Dead (Coleman et al., 2007) focuses more on post-mortem events; various versions of translations will be cited profusely in Chapters 5, 6. But this book itself draws on Hindu ideas of human anthropology. Hence the choice of sources – is a popular presentation of anthropological data (Hinduism. Anthropology, n / d; Satpathy, 2018) and scientific presentation (Dasgupta, 1922; Radhakrishnan, 2009/1923).

BEFORE THE WRITING HAD EMERGED

In artifacts and texts, there are certain ideas about the posthumous existence of that part of a person that does not die with the body. This part is described as spirit, soul, Ib, atma, etc., and these terms have different content. Therefore, it is imperative that before reconstructions and comparisons of post-mortem scenarios (Chapters 5, 6) are carried out, all these terms must be compared.

Coordinate system. All terms should be compared with the "coordinate system" for a correct comparison. As such a system, we use the human-environment model. Thus, a basis will be created for reconstructing and comparing post-mortem scenarios.

Upper Paleolithic conceptions. Suppose Figures 4 and 5 (Chapter 3) are understood correctly. In that case, the connection between the ideas of posthumous existence and the cult of fertility exists. Rebirth is natural just like the appearance of foliage on a tree in the spring after their winter absence (image of the World Tree, Figure 1, Chapter 3), and occurs in a generation. The comparison of concepts will be shown in Table 1. Naturally, the names of the mortal and posthumous components have not been preserved, and they can only be judged from the surviving images.

Table 1. Comparison of the Upper Paleolithic concepts with the SWM of a person

Concept	Main attributes	Correspondence to the soliton-wave model
Body	Dies, motionless	Soliton component
Soul-bird	Does not die, flies out of the body like a bird, returns to the body of a grandson or granddaughter	Wave component, subjective space as a whole

Mesolithic and Neolithic views. More scenes of violence and death appear in cave and rock art; the images themselves become more schematic, contrasting ("figure" and "background"), and dynamic. The activity of the ego, will, and reason does not leave a place for the Paleolithic ideas about the afterlife. Instead of passively waiting for rebirth, humans must actively fight for it! Perhaps it went something like this:

A person who was vitally important to follow the path of the dead descended into the valley and crossed the river. Then he was subjected to ritual sacrifice and became dead "in fact." What happened next was in a state of trance, when illusions are easily mistaken for reality. The person tried to deceive the underworld rulers by wearing a mask or a chimerical costume (Figure 1).

Figure 1. Disguised shaman. National Historical and Archaeological Reserve "Stone Grave". Ukraine, Melitopol

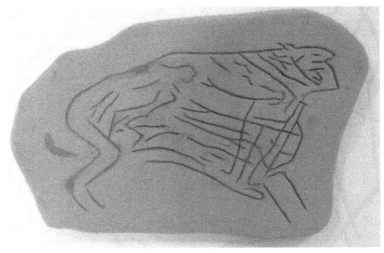

(Y. N. Kovalyov, with the permission of the museum staff)

Perhaps he was trying to defeat them by force. As a sign of his victory, he planted his foot on the head of a killed snake (Figure 2). He achieved immortality in the world of higher spirits, completing his journey through the worlds of the World Tree.

Figure 2. Head of a snake with a foot sign. National Historical and Archaeological Reserve "Stone Grave". Ukraine, Melitopol

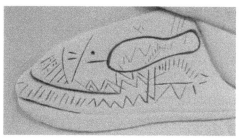

(Y.N. Kovalyov, with the permission of the museum staff)

The location in the grotto of the Stone Tomb - the oldest object associated with the cult of the dead - leaves no doubt: the shaman ensures his safety when meeting with the inhabitants of the afterlife.

Let us summarize this reasoning in Table 2.

Table 2. Comparison of Mesolithic and Neolithic concepts with the SWM of a person

Concept	Main attributes	Correspondence to the soliton-wave model
Twin-soul	Purposefulness, will, and readiness to fight by all means	Wave component, subjective space as a whole; levels 2 and 3 are active

Comparison with the previous period shows an important circumstance: the posthumous consciousness of a person does not remain unchanged from epoch to epoch, it evolves along with the evolution of a living person.

ENEOLITHIC IDEAS, SUMERIAN-AKKADIAN MYTHS, EGYPTIAN BOOK OF THE DEAD

The combination of two different sources here is justified by the fact that they are relatively synchronous and created in similar conditions: for example, the cultures of Mohenjo-Daro and Cucuteni-Trypillia (Korvin-Piotrovsky, 2008) were Eneolithic and agricultural, like the culture of the Sumerians; both build a large settlement, having all the signs of cities and have a complex social organization, a certain religion, the beginnings of writing. In addition, images of Trypillian ceramics, seals, and symbols of Mohenjo-Daro are supplemented and, as it were, commented on by Sumerian myths.

What does such a comparison show?

Firstly, it preserves Neolithic heritage, but in a revised form. Direct intuitive sensation is no longer available, there is no clear knowledge of the world of the dead, and it is depicted in the lower part of Figure 3 as something indefinite, chaotic, in need of additional explanation, located underground.

Secondly, the fear of the afterlife intensifies: mere mortals can no longer pass it - only the goddess Inanna was able to return, giving the soul of her lover as a ransom, only the great hero and two-thirds of the god Gilgamesh, using the special patronage of Shamash and equipped with the necessary amulets were able to return alive, only Uta-napishtim (Ziusudra), who had earned the mercy of the gods and survived the Deluge, was able to find eternal life there.

Figure 3. World tree

(Albedil, 1991)

Thirdly, here are descriptions of the soul state in the afterlife:

1. Inanna's descent into the Netherworld (Dedović, 2020):

[129-133] And when Inanna entered the first gate, the turban, headgear for the open country, was removed from her head. "What is this?" "Be satisfied, Inanna, a divine power of the underworld has been fulfilled. Inanna, you must not open your mouth against the rites of the underworld."

[134-138] When she entered the second gate, the small lapis-lazuli beads were removed from her neck. "What is this?" "Be satisfied, Inanna, a divine power of the underworld has been fulfilled. Inanna, you must not open your mouth against the rites of the underworld."

[139-143] When she entered the third gate, the twin egg-shaped beads were removed from her breast. "What is this?" "Be satisfied, Inanna, a divine power of the underworld has been fulfilled. Inanna, you must not open your mouth against the rites of the underworld."

[144-148] When she entered the fourth gate, the "Come, man, come" pectoral was removed from her breast. "What is this?" "Be satisfied, Inanna, a divine power of the underworld has been fulfilled. Inanna, you must not open your mouth against the rites of the underworld."

[149-153] When she entered the fifth gate, the golden ring was removed from her hand. "What is this?" "Be satisfied, Inanna, a divine power of the underworld has been fulfilled. Inana, you must not open your mouth against the rites of the underworld."

[154-158] When she entered the sixth gate, the lapis-lazuli measuring rod and measuring line were removed from her hand. "What is this?" "Be satisfied, Inanna, a divine power of the underworld has been fulfilled. Inanna, you must not open your mouth against the rites of the underworld."

[159-163] When she entered the seventh gate, the pala dress, the garment of ladyship, was removed from her body. "What is this?" "Be satisfied, Inanna, a divine power of the underworld has been fulfilled. Inanna, you must not open your mouth against the rites of the underworld."

[164-172] After she had crouched down and had her clothes removed, they were carried away. Then she made her sister Erec-ki-gala rise from her throne, and instead, she sat on her throne. The Anuna, the seven judges, rendered their decision against her. They looked at her -- it was the look of death. They spoke to her -- it was the speech of anger. They shouted at her -- it was the shout of heavy guilt. The afflicted woman was turned into a corpse. And the corpse was hung on a hook.

2. The Epic of Gilgamesh (George, 2016) [Gilgamesh asks the spirit of Enkidu, who returned from the underworld for a while]:

'O tell me, my friend! Tell me, my friend!
Tell me what you saw of the ways of the Netherworld!'
'[I, the] friend whom you touched so your heart rejoiced,

[my body like an] old garment the lice devour.
[Enkidu, the friend whom you] touched so your heart rejoiced,
[like a crack in the ground] is filled with dust.'
'Did [you see the man with one son?]' 'I saw him.
[A peg is] fixed [in his wall] and he weeps over [it bitterly.],
'[Did you see the man with two sons?' 'I] saw him.
[Seated on two bricks] he eats a bread-loaf.'
'[Did you see the man with three sons?]' 'I saw him.
He drinks water [from the waterskin slung on the saddle.]'
'Did [you see the man with four sons?]' 'I saw him.
[Like a man with a donkey]-team his heart rejoices.'
'Did you see [the man with five sons?]' 'I saw him.
[Like a] fine [scribe] his hand is nimble,
he enters the palace [with ease.]'
'Did you see [the man with six sons?]' 'I saw him.
[Like a ploughman his heart rejoices.]'
'[Did you see the man with seven sons?' 'I saw him.]
[Among the junior deities he sits on a throne and listens to the proceedings.']
['Did you see the one with no heir?' 'I saw him.]
He eats a bread-loaf like a kiln-fired brick. ']
['Did you see the palace eunuch?' 'I saw him.]
Like a fine standard he is propped in the corner, like .... '
'Did you see the one who was struck by a mooring-pole?' 'I [saw him.]
Alas for his mother [and father!] When pegs are pulled out [he]wanders about.'
'Did you see the one who [died a] premature death?' '[I saw him.]
He lies on a bed drinking clean water.'
'Did you see the one who was killed in battle?' 'I [saw him.]
His father and mother honour his memory and his wife [weeps]over [him.]'
'Did you see the one whose corpse was left lying on the plain?' 'I saw him.
His shade is not at rest in the Netherworld.'
'Did you see the one whose shade has no one to make funerary offerings?'
'I saw him.
He eats scrapings from the pot and crusts of bread thrown away in the street.'
XII 90-150 OB version

So, there are two distinct descriptions; let's try to find invariants related to the post-mortem state.

There are four of them: this is 1) the preservation of the personality, 2) its passive final state, 3) the loss of certain attributes (7 for Inanna, 1 body for Enkidu), leading to such a state, 4) deprivation of attributes as a result of the laws of the underworld and a certain court, repaying a person according to his deeds (The Descent of Inanna says this directly, the words of Enkidu simply mention the consequences; Inanna wanted to unite both worlds under her rule, and was punished by deprivation of all attributes and turning into a corpse hanging on a hook; in the case of ordinary mortals, they part only with the body).

Thirdly, in the Descent of Inanna, attributes are spoken of as some kind of magical amulets. In the Epic of Gilgamesh there is this story [Gilgamesh and Enkidu kill a creature called Humbaba]:

> Gilgamesh, having heard the word of a friend,
> Again, he relied on his strength:
> "Hurry, come closer, so that he does not leave us,
> If he hadn't gone into the forest, he wouldn't have hidden from us.
> He dresses in seven terrible robes,
> He put on one and took off six more.

(Table 4)

> Gilgamesh tells him, Enkidu:
> "When we come to slay Humbaba,
> Rays of radiance in confusion will disappear,
> Rays of radiance will disappear, the light will be eclipsed,
> Enkidu tells him, Gilgamesh:
> "My friend, catch the bird - the chickens will not leave!
> We'll look for rays of radiance later,
> Like chickens in the grass, they will run away
> Kill yourself, and the servants later.
> When Gilgamesh heard his comrade's word,
> He raised his battle ax
> He drew his sword from his belt,
> Gilgamesh struck him in the back of the head,
> His friend, Enkidu, struck him in the chest,
> On the third blow he fell,
> His violent members froze.
> They struck down the guard, Humbaba, -
> Cedars groaned for two fields around,
> With him Enkidu slew the forests and cedars,

Enkidu slew the guardian of the forest,
Whose word Lebanon and Saria feared.
Peace embraced the high mountains,
Peace embraced the wooded peaks,
He slew the keepers of the cedars -
Broken beams of Humbaba.
When he killed all seven of them,
A battle net and a dagger of seven talents, -
A load of eight talents - removed from his body,
He opened the secret dwelling of the Anuna.

(Table 5)

Here again, some 7 attributes are mentioned - robes, rays - which must be destroyed for the final death of Humbaba, and these rays and robes belong to his personality and are relatively autonomous.

In both stories, one can see a hint of the nature of these attributes - these are 7 vital shells that are components of the system-personality. If this assumption is correct, then for true death it is necessary to first lose all its constituent parts, while "ordinary" death deprives only the body.

It is difficult to restore the functions of each of the shells, which would give the key to correlation with the levels and channels of a person - both texts are very dark (intentionally, or because the true meaning of the attributes has already been forgotten). However, let's try to do this based on the type and location of Inanna's amulets (according to (Khazarzar, n / d), as well as the correlation of body parts with elements[1]:

- Crown of Eden - head, spirit - with intuition.
- Signs of dominion and judgment - with the ego.
- Necklace - with reason will.
- Pendant on the chest - with heart and mind (nous).
- Wrists on hands - with action, energy.
- Grid "To me, man, to me" - with sexual passion (generally, with feelings and emotions).
- Bandage on the hips - with the ground (generally, with the body).

The image of the Tree of Life (Figure 4) can provide additional data:
- Its symmetry is emphasized, even though different gods correspond to the symmetrical halves, which can be compared with the oblique symmetry of the wave and soliton components.
- Openness (cultivation by gods, kings, or priests).

- Hierarchy, the same for symmetrical halves; sometimes there are 6 such branches or flowers and 1 on top, but other quantities are also found.

Figure 4. Ashurnasirpal II performs religious rituals before the sacred tree. From Nimrud, Iraq. 865-860 BCE. British Museum

(O. S. M. Amin, 2016)

Let's summarize the obtained comparisons in Table 3, prefixing them with images of the heroes of myths (Figure 5).

Figure 5. Heroes of Sumerian-Akkadian myths 1 – Inanna. (Sailko, n / d); 2 – Erec-ki-gala. (BabelStone (n / d); 3 – Gilgamesh. (U0045269, n / d); 4 – Enkidu (O. S. M. Amin, n / d)

Table 3. Comparison of Eneolithic ideas with the soliton-wave model of a person

Concept	Amulets of Inanna	Compliance with the soliton-wave model
Body	Thigh bandage	The solution part of the system
Spirit:	All other attributes:	The wave part of the system
1) feelings, passions	Grid "To me, man, to me"	Channels 5, and 6th levels
2) actions, energy	Wrists on hands	Potential, modality of actual existence
3) nous	Pendant on the chest	Nous is a channel of the 4th level
4) reason, will	Necklace	Reason and will - channels of the 3rd level
5) ego	Signs of Dominion and Judgment	Ego is a 2nd level channel
6) spirit, head	Crown of Eden	Intuition - Level 1 Channel

continued on following page

Table 3. Continued

Concept	Amulets of Inanna	Compliance with the soliton-wave model
7) the basis of personality	Corpse on a hook	The modality of the potential existence of the ego
Death	1. Destruction of the system of spirit and body. 2. Neti, the main guardian of the kingdom of the dead, producing the destruction of the system of soul and body	1. Destruction of the soliton-wave system. 2. The potential required to destroy the multilevel bonds of the components of the soliton-wave system, that is, a kind of "anti-system"

If the interpretation made is correct, there are very mature ideas about the structure of human subjective space. Let's look at the personification of death. Let's compare these ideas with ancient Egyptian ones.

The work of the subjective space structure reconstruction is facilitated by the correlations of the names of structures with their functions, available in publications. Let us cite, somewhat shortened, two such texts. According to (Shaposhnikov, 2003), we can build such a scheme (Figure 6):

Figure 6. Ancient Egyptian ideas about the structure of humans according to (Shaposhnikov, 2003)

1. Khet – material body	The main purpose of Khet is interaction with the material world and actions in it. The Egyptians valued the safety of the body of the deceased, most of all, about the head as the "seat of life." Decapitation and burning were considered terrible fates. In order to insure the deceased in case of the destruction of Khet, the Egyptians installed portrait copies in the tombs, into which his energy shells could be infused. The gods also have Khet, a body given in sensations. In addition to the natural bodies, the gods began to use shells made by people - statues, sacred objects, and images.	
2. Ku or Ka, Ke – vital essence, ethereal body, human energy double, soul-double	Ka is a set of a person's mental sensations, and it is associated with the personality, individuality, and bodily and spiritual features of the deceased. Initiates could see Ka as a rainbow radiance around a material body. The material body and the energy double of a person are not separated, but with poor health, nervous shock, or excitement, Ka can partially leave Khet, and a person falls into a semi-conscious state or trance. Ka lives in the tomb, and there accept energy counterparts of offerings from living relatives. The gods also had Ka. Ptah had his Ka in the sanctuary of Memphis; Ra had 14 Ka - male and female aspects of the Sun, Earth, Moon, Mercury, Venus, Mars, Jupiter, and Saturn.	
3. Bi or Ba, Be – essence of a person, "life force", the soul-manifestation, the shell of the subconscious, the "astral body"	Ba is formed from the human feelings, desires, emotions. Ba with amazing speed changes its form under the influence of each impact of sensations, feelings, desires and thoughts. In the Old Kingdom, it was believed that only gods, kings and high priests possessed Ba. Ba was conceived as something separate existing only after the death of a great initiate. It was also believed that Ba is an energy that animates a god statue or fetish of, or a mummy. When Ba separates from Khet, the latter falls into a sleepy stupor. The initiated Egyptians could, at their will, perform in the form of a journey to various places and to the other world. At the same time, Ba, which, like a bird, could leave the body of a sleeping person, a mummy in a tomb, a statue of a god or a king and move as far away as desired, invariably had to return to the body whose soul it was. Ba was depicted as a falcon with a human head and was sometimes depicted sitting on a tree near the tomb, drinking water from the pond, but without fail descending into the tomb to the body with which it was associated. Ba form the world of other souls and the world of dreams. The Ba of the deceased had the ability to move into other bodies, to pass into another material entity (in the divine golden hawk, in the Phoenix bird, in the crane, swallow, ram, crocodile, snake). The gods also had Ba, often several. God Pa possessed the Ba family, the astral energies of the Sun and Earth, the Moon, Mercury, Venus, Mars, Jupiter and Saturn. In addition, the planet Mars was considered Ba Hora, Jupiter - Ba Hora and Ba Seta, Saturn - Ba Horus bull. The fixed stars and constellations were also regarded as the Ba of the gods: the constellation of Orion was considered the Ba of Osiris (especially Orion's Belt), the constellation Canis Major (the star Sirius) was considered the Ba of Isis. Sometimes one God is referred to in the texts as Ba Nun, Apis - Ba Ptah, Sokaris - Ba Osiris, etc.	
4. Ib or Eb - the soul-heart, the receptacle of human consciousness, the "mental body"	Ib is an immortal, extremely mobile, transparent and gentle soul, formed by human thoughts and mental images. According to the feelings of the initiates, Ib acquires a radiant unearthly beauty. The initiates considered the heart to be the center of human consciousness. Hence - the single naming of the "mental body" and "heart". After death, Ib returns to his universal source - Ib of the god Osiris. Ib was considered as something aware of the hidden thoughts of a person and the secret motives of his actions. Therefore, at the Afterlife Court, Ib could become a dangerous witness, and the "Book of the Dead" (Chapters 27, 30) contains spells that encourage Ib not to testify against the deceased. In the process of mummification of the body, they often put an artificial heart in the form of a scarab statue with spells inscribed on it. Ib-scarab was supposed to provide the deceased with favorable evidence about his earthly deeds. This symbolism describes Ib as the energy of the Sun, because the scarab is a symbol of the god of the rising Sun Khepri (incarnations of Ra).	
5. Ib or Eb – the soul-cause or super consciousness, "karmic body"	The soul-cause is immortal, transmits information to the next incarnations in the form of unconscious aspirations. It is responsible for the place and time of a person's birth, all his innate bodily vices and diseases. It is the soul-cause that allows a person to be born in a certain family, clan, tribe, and so on, with which she had connections in previous incarnations.	
6. Ib or Eb – the soul-meaning or self-consciousness	The soul that produces meaning, thanks to it, a person can observe the course of his own thoughts, realize his existence, see the innermost meaning of his life. The Ib (consciousness) is polluted by evil mental images, then they prevent the soul-sense (self-consciousness) from perceiving the infinity of consciousness, just as clouds and darkness prevent the Sun from perceiving the surface of the Earth.	
7. Ah or Sah - the "spiritual body", the part of the universal energy sub-foundation of the Universe	Ah is immortal, boundless, permeates everything that exists in the universe. Ah means "bright, enlightened, illuminated, blissful". Ah at every point in space and contains all the information in all its forms. Ah resides both in the material world and in the incorporeal world, he is omnipresent. Ah, one for all. This spirit protects from evil thoughts, words and deeds, blocking its source with dense barriers of the causal shell. Ah is also among the gods. Most often it is mentioned, Ah of Osiris, Horus, Ra, as well as a collective multiplicity of spirits-souls of otherness, which hospitably or hostilely meet his Ka, Ba, Ah of the deceased. Ah was depicted as a crested ibis.	

According to (Granin, 2011): "The physical body of a person is Khet ("what decay is inherent in"), its material preservation through mummification was a necessary condition for the maturation of a spiritual body (Sah) in it.

The center of the spiritual life and thinking of a person, the source of animal strength, the focus of the will is the heart - Ib. It was the body in which the vices and virtues of a person were localized - a complex called conscience. For this reason, at the trial of Osiris, the heart was weighed on the scales of Isis and Nephthys. Since the test was spiritual, it was not so much the material heart that was tested, but its spiritual correlate Ba ("majestic", "noble", "powerful") - the soul of the heart or the soul of the human person, while the heart was more of an iconographic symbol.

Ba is an element of the divine mind in man, the bearer of higher consciousness, and the receptacle of pure ideas. Ba carried out the transcendental function of transition from the material form of existence to the spiritual one. In the physical dimension, she lived in a tomb. In the "astral", or "ethereal" form, she lived in the afterlife with Ra and Osiris, eating divine food. The function of Ba could be performed by one God to another. So, Ra was Ba Nun, the deity of the primordial water chaos, from which the sun god Atum-Ra arose. This plot allows us to interpret Ba as a symbol of the birth of consciousness by separating it from the unconscious. Also, stars can act as Ba (Ba of Osiris is Orion), the Bennu bird, and so on. The soul of Ba possessed the shadow of Khaibit, which, apparently, is an archaic relic, when his shadow was considered a person's soul. In functional and structural terms, we consider it identical to Ba.

The counterpart of the personality in its vital state is the shadow - Shut, later Ka, or a person's "energetic" counterpart. The presence of this concept testifies in favor of the existence among the Egyptians of a fundamental ontological problem of the relationship between the material and the spiritual. Ka answers the question of how the spiritual (Ba) can relate to the material. Ka, like Ba, performs a mediumistic (transcendental) function, it connects the soul with the body, while not being fully either matter (even though it needs material food in the form of funeral offerings) or spirit. Ka is ambivalent (occultists would call it "the astral body").

Ka is the link constituting the elements of the human personality (Khet, Ib, Ba) into a single individual. Ka is also a guarantor of the preservation of personality after death, acting as a kind of matrix consisting of Khet (the deceased physical body), Sah (spiritual body ripening in Khat), and Ba (through the energy of which Sah ripens). On the same level with Ka in importance is the Name - Ren, the sacred symbol of personality, since the spirit is the name, and the name is the spirit, and the destruction of the name entails the destruction of man. Sah is the non-material, "ethereal", spirit-bearing body of the immortal soul Hu. Sah's body dwells among the gods. It is the form of the god bestowed by the deceased Osiris. In this sense, it was called the Sah of Osiris. Hu ("bright", "glorious") is the immortal soul of a person belonging to the divine world. She is endowed with the embodiment of the life energy Sekhem. Initially, the term "hu" denoted a class of celestial beings who lived among the gods. Thus, Hu is transcendent to a person during his physical life and appears as a part of the personality only after death and after a new spiritual body Sah, absent during life, matures in him. According to the exoteric tradition, Sah matured from the physical body (Khet) of the deceased with the appropriate performance of the funeral ritual. According to esoteric ideas, the formation of Sah requires not so much a physical body as its counterpart Ka, which is attuned to the divine soul-consciousness Ba, as a result of which a powerful influx of energy into the Ka body occurs, after which a critical mass is reached, and Ka flares up with an ethereal flame, transforming into the spiritual enlightened body of Sah[2]. The function of Ba in the post-mortem state is its correspondence to transcendental reality (symbolic weighing of the heart). In the above representations, the idea is traced that any property has its carrier, and any carrier has a certain "substantial" property, i.e., there is a dichotomy of the "body/soul" type. The carrier belongs to one level of reality, and its embodied property belongs to a higher level. So, for example, the material (Khet) and spiritual (Sah) bodies need correlates of the highest level - in the soul of the heart (body) Ba and the soul of the spirit Hu. Thus, a correlation is found between the levels of the human personality and the stratification of the levels of reality. Death is a kind of line of symmetry between man and "external" reality.

Comparing these interpretations, one can state their significant differences, the smallest of which is vowels (the ancient Egyptian language is dead; attempts to restore vowels based on modern Coptic pronunciation will always cause controversy).

It is much worse than references to the occultists and terms like "astral body" or "mental body", coined by H.P. Blavatskaya, are used for interpretation., which is permissible only out of complete desperation. This is a paradigmatic problem of modern science, which does not find suitable concepts and means for the exegesis of ancient texts, as well as an example showing the difference between the consciousness of the ancient Egyptians and the consciousness of modern man.

However, the same problem of paradigmatic insufficiency faced the ancient Egyptians, who tried to convey intuitive knowledge about the world of the dead in their era means, when the channel of intuition was already greatly weakened, and the channel of the mind was not yet sufficiently developed (Kovalyov et al., 2020).

To solve this problem, they used an apparatus that resembles what is described in Chapter 12. So, there is a similarity with the ratio of invariants and gauges (the god Ra is an invariant, and his Ba (Sun, Earth, Moon, Mercury, Venus, Mars, Jupiter, and Saturn) – calibrations, as in the quoted fragment). There is also a similarity with ternary connectives: Ra is the object of intuitive knowledge; its Ba are projections that provide the reader - the subject - with the intuitive perception of Ra. There are symbols (images, names, plots of individual myths) and signs. Multiple interpretations of the same symbol in the fragment of the dialogue part of the Book of the Dead cited above distract the reader from fixing the symbol as a sign with one connotation and show that the symbol is a kind of "direction" towards intuitive knowledge, limited by "frames", directly forming a symbol. Also, different shells' hierarchy, relationships, evolution, and external interactions are like an open system.

Let's apply this apparatus and reconstruct the Egyptian ideas about the structure of man.

The universal spirit Ah (Hu) contains the potential foundations of personality, which are brought into actual existence by Ren (spell). A pair of Shut-Khet (later Ka-Khet) is formed, with the "thin" side of the Shut facing Ah and the "rough" side of Khet facing the material world. Thus, the Shut-Khet pair can be considered as the boundary space of the ternary connection between matter and spirit. Ib structures correlating with the systems and organs of the body have the same pair of characteristics. Ka (later accommodating the Shut functions) represents sensations and emotions and has a correlation of the sense organs of the body.

Ba is a projection of a god into the material world (as Ba Ra are the Sun, Moon, and planets), or into the non-material world of the gods (as Ra is Ba Nun), or the dead into the material world (as the Ba of the deceased is the golden hawk, in phoenix bird, crane, swallow, ram, crocodile, snake), or the great gods initiated into the world.

Sah is the body of the gods who died in the world, bestowed by Osiris (the Ba of the deceased receives Ah from Osiris).

The ratios of the shells of a human with the HEM channels and levels are shown in Table 4.

Table 4. Comparison of ancient Egyptian ideas with the soliton-wave model of a person

The concept	Shell	Correspondence to the soliton-wave model
Universal spirit	Ah (Hu)	Potential, modality of potential existence, intuition - channel of the 1st level
Person's name	Ren	Ego actualization
Shadow	Shut (later Ka)	Wave part of the system
Physical body	Khet	Soliton part of the system
Self-consciousness	Ib - 3	Ego is a 2nd level channel
Motives	Ib - 2	Reason and will - channels of the 3rd level
Mind	Ib - 1	Nous - a channel of the 4th level
Sensations, vital energy, biofield	Ka	Sensations - channels 5, and 6 levels (later the wave part of the system)
External manifestations	Ba	Projection
Spiritual body bestowed by Osiris	Sah (Ba+Ah)	Soliton-wave systems in the god's world

Both in the Egyptian and Sumerian traditions we have an idea about the structure and attributes of the system of shells (amulets), its openness, and evolution. The Egyptians perceive this evolution positively - in the end, the deceased (pharaoh, high priest, or initiate), thanks to the grace of the lord of the underworld, Osiris, acquires a spiritual body and becomes able to live among the gods (mere mortals continue to exist, like earthly). The evolution of the Sumerians is strictly negative - Inanna is deprived of all amulets and, at the behest of the mistress of the underworld, Erec-ki-gala, turns into a corpse hanging on a hook (mere mortals who do not infringe on power in the afterlife continue to exist similar to the earthly one).

If compared with the soliton-wave system, then uncertainties are visible: there is no clear separation of gauges and modalities from the structure, the potential is not defined (in the Sumerian myth), the emergence and openness of the system are not emphasized, if the scenario of convolution with a decrease in potential is seen in the descent of Inanna, then the acquisition Sah's body is described too vaguely. These ambiguities are determined by the level of activity of channels of subjective space during the Eneolithic period and the first civilizations.

If we talk about the epistemological side, in both cases, a prototype of the ternary connective is used, which was greatly facilitated by the magical consciousness of the Sumerians and the Egyptians.

As consciousness evolved and channel activity changed (and the Sumerian and Egyptian myths lasted long enough for the changes to become noticeable), the symbols and plots became more and more difficult to perceive. Moreover, suppose the fate after death was ultimately determined by the wrath/mercy of the gods. In that case, the desire to discard "all these difficulties" and focus on gaining divine mercy is understandable. Hence the development of the soteriological idea in subsequent religious experience, which acquired completed forms in the Abrahamic religions.

ABRAHAMIC RELIGIONS

The structure of man is defined as follows: a mortal body and an immortal soul, or body, soul, and spirit. Let's determine more precisely the ratio of spirit, soul, and body, comparing them with the coordinate system of the soliton-wave model.

St. Gregory of Nyssa writes: *"Because our nature is dual: one is subtle, spiritual and light, and the other is thick, material and heavy"* (St. Gregory of Nyssa, 1862). In Holy Scripture, the body is defined as the "house", "tabernacle" and "clothes" of the soul, i.e., like its shell. Thus, the soul and body can be compared with the wave and soliton components of the soliton-wave model, but the body is also spoken of as a shell of the soul.

On the other hand, it is also said that the soul's house is the body, and the spirit's home is the soul. Ep. Theophan, the Recluse, explains: *"In a person, one must distinguish between soul and spirit. The spirit contains the feeling of the Divine: conscience and unsatisfactoriness with anything. He is the power breathed into the face of man at creation. The soul is a lower force or a part of the same force assigned to conduct the affairs of earthly life. She is such a force as the soul of animals but exalted to combine the Spirit with her. The spirit of God, combined with the soul of animals, elevated it to the level of the human soul. The Spirit, as a force that has come from God, knows God, seeks God, and finds rest in Him alone. Convinced by some spiritual innermost instinct of his origin from God, he feels his complete dependence on Him. He recognizes himself obliged to please Him in every possible way and live only for Him and Them"* "(St. Theophan the Recluse, 2013).

Or, according to the explanations of modern commentators: *"It is most correct to understand that the soul and spirit of a person are not two separate entities, but two sides of the same spirit, two parts in one whole. For if we admit that the soul and the spirit are separate entities, then it must be admitted that, after the death of a person, they must exist separately from each other, or else the lower part must be destroyed, as happens with the soul of animals. Still, such a thought is unacceptable because both Holy Scripture and all St. The Fathers of the Church unanimously speak of the existence and complete preservation of a single soul beyond the grave"* (Human tripartiality, n / d).

Therefore, the soul is a kind of shell of the spirit; there is a certain hierarchy of qualities of spirit and soul; both the spirit and the soul are united in their quality of "lightness", as opposed to the "thickness" of the body and can be compared with different levels and channels of the wave component.

To clarify with what levels and channels, we will consider the creation, fall, and expulsion of man from paradise, citing the text in full (Holy Bible – ASV., n / d).

27 And God created man in his own image, in the image of God created he him; male and female created he them.

28 And God blessed them: and God said unto them, Be fruitful, and multiply, and replenish the earth, and subdue it; and have dominion over the fish of the sea, and over the birds of the heavens, and over every living thing that moveth upon the earth.

29 And God said, Behold, I have given you every herb yielding seed, which is upon the face of all the earth, and every tree, in which is the fruit of a tree yielding seed; to you, it shall be for food:

(Gen. 1)

15 And Jehovah God took the man and put him into the garden of Eden to dress it and to keep it.

16 And Jehovah God commanded the man, saying, Of every tree of the garden thou mayest freely eat:

17 but of the tree of the knowledge of good and evil, thou shalt not eat of it: for in the day that thou eatest thereof thou shalt surely die.

18 And Jehovah God said, It is not good that the man should be alone; I will make him a help meet for him.

19 And out of the ground Jehovah God formed every beast of the field, and every bird of the heavens; and brought them unto the man to see what he would call them: and whatsoever the man called every living creature, that was the name thereof.

20 And the man gave names to all cattle, and to the birds of the heavens, and to every beast of the field; but for man there was not found a help meet for him.

21 And Jehovah God caused a deep sleep to fall upon the man, and he slept; and he took one of his ribs, and closed up the flesh instead thereof:

22 and the rib, which Jehovah God had taken from the man, made he a woman, and brought her unto the man.

23 And the man said, This is now bone of my bones, and flesh of my flesh: she shall be called Woman, because she was taken out of Man.

24 Therefore shall a man leave his father and his mother, and shall cleave unto his wife: and they shall be one flesh.

25 And they were both naked, the man and his wife, and were not ashamed.
(Gen. 2)

1 Now the serpent was more subtle than any beast of the field which Jehovah God had made. And he said unto the woman, Yea, hath God said, Ye shall not eat of any tree of the garden?

2 And the woman said unto the serpent, Of the fruit of the trees of the garden we may eat:

3 but of the fruit of the tree which is in the midst of the garden, God hath said, Ye shall not eat of it, neither shall ye touch it, lest ye die.

4 And the serpent said unto the woman, Ye shall not surely die:

5 for God doth know that in the day ye eat thereof, then your eyes shall be opened, and ye shall be as God, knowing good and evil.

6 And when the woman saw that the tree was good for food, and that it was a delight to the eyes, and that the tree was to be desired to make one wise, she took of the fruit thereof, and did eat; and she gave also unto her husband with her, and he did eat.

7 And the eyes of them both were opened, and they knew that they were naked, and they sewed fig-leaves together, and made themselves aprons.

8 And they heard the voice of Jehovah God walking in the garden in the cool of the day: and the man and his wife hid themselves from the presence of Jehovah God amongst the trees of the garden.

9 And Jehovah God called unto the man, and said unto him, Where art thou?

10 And he said, I heard thy voice in the garden, and I was afraid, because I was naked; and I hid myself.

11 And he said, Who told thee that thou wast naked? Hast thou eaten of the tree, whereof I commanded thee that thou shouldest not eat?

12 And the man said, The woman whom thou gavest to be with me, she gave me of the tree, and I did eat.

13 And Jehovah God said unto the woman, What is this thou hast done? And the woman said, The serpent beguiled me, and I did eat.

14 And Jehovah God said unto the serpent, Because thou hast done this, cursed art thou above all cattle, and above every beast of the field; upon thy belly shalt thou go, and dust shalt thou eat all the days of thy life:

15 and I will put enmity between thee and the woman, and between thy seed and her seed: he shall bruise thy head, and thou shalt bruise his heel.

16 Unto the woman he said, I will greatly multiply thy pain and thy conception; in pain thou shalt bring forth children; and thy desire shall be to thy husband, and he shall rule over thee.

17 And unto Adam he said, Because thou hast hearkened unto the voice of thy wife, and hast eaten of the tree, of which I commanded thee, saying, Thou shalt not eat of it: cursed is the ground for thy sake; in toil shalt thou eat of it all the days of thy life;

18 thorns also and thistles shall it bring forth to thee; and thou shalt eat the herb of the field;

19 in the sweat of thy face shalt thou eat bread, till thou return unto the ground; for out of it wast thou taken: for dust thou art, and unto dust shalt thou return.

20 And the man called his wife's name Eve; because she was the mother of all living.

21 And Jehovah God made for Adam and for his wife coats of skins, and clothed them.

22 And Jehovah God said, Behold, the man is become as one of us, to know good and evil; and now, lest he put forth his hand, and take also of the tree of life, and eat, and live for ever-

23 therefore Jehovah God sent him forth from the garden of Eden, to till the ground from whence he was taken.

24 So he drove out the man, and he placed at the east of the garden of Eden the Cherubim, and the flame of a sword which turned every way, to keep the way of the tree of life.

(Gen. 3)

So, before the fall, a person has:

1. Intuition (communicates with God, sees the essence of all animals and birds, giving them the appropriate names, not knowing how Eve was created because he slept, he recognizes this upon awakening).
2. Ego (separates itself from God, Eve, animals, and birds).

3. Reason (he could listen to God, or a snake, judging that it was better, but did not do this) and free will (he was able to violate the commandment, figuratively speaking, in front of God, and did not repent).
4. Nous or practical mind *(And when the woman saw that the tree was good for food, and that it was a delight to the eyes, and that the tree was to be desired to make one wise, she took of the fruit thereof, and did eat; and she gave also unto her husband with her, and he did eat)*, which turned out to be stronger than reason.

Speaking about the soul in its current state, commentators of the Holy Scriptures define its qualities as follows: *"verbal, irritable, lustful"* and explain *"The difference between the soul and the spirit is also evident from the fact that a person who lives according to the inclinations of sensual flesh and his animal strength, sometimes nothing does not differ from dumb animals, but even becomes worse than them, not knowing any boundaries and limits, while animals do not cross these boundaries, being content with only the necessary. Such people are usually called carnal, or spiritual. Those who live according to the inclinations of a God-like spirit are called spiritual."* (Human tripartiality, n / d).

Thus, the result of the fall was the strongest "clouding" - the weakening of intuition, ego, reason, and will, and the development of a nous (4th level) and feelings (5-6th levels), as if subjugating spiritual qualities.

We also note two types of death: *"For just as the separation of the soul from the body is the death of the body, so the separation of God from the soul is the death of the soul. And this is mainly death, the death of the soul. God pointed to her, and when, giving a commandment in paradise, he said to Adam: "On what day you eat from the forbidden tree, you will die by death" (Gen. 2, 17)"* (St. Palamas, 2004).

This "death of the soul" was the fall into sin – and here it is appropriate to compare it with the unfolding scenario when the potential of the higher levels is partially spent on the formation of the lower ones and is not compensated from the outside – "there is no grace". Such a comparison is also supported by indications of the unevenness of the passage of time, which is repeatedly mentioned in the New Testament, for example:

8 But forget not this one thing, beloved, that one day is with the Lord as a thousand years, and a thousand years as one day.

(2 Peter 3)

The immortality of a person in paradise can be understood both as the absence of time there (1-3 levels, Chapter 1), and as his appearance, after being expelled from paradise (level 4), and acceleration after the Flood (time accelerates at each subsequent level).

The idea of the original immortality of humans characterizes the Orthodox tradition. Only the sin committed by Adam and Eve voluntarily, in which they did not immediately repent, introduced the mortal principle into the essence of humans:

13 God did not create death and does not rejoice in the destruction of the living,

14 for He created everything for existence, and everything in the world is saving, and there is no destructive poison, there is no kingdom of hell on earth.

15 Righteousness is immortal, but unrighteousness causes death:

16 The wicked attracted her with both hands and words, considered her a friend and withered away, and made an alliance with her, for they are worthy to be her lot.

(Book Wisdom of the Solomon, 1); missing from (Holy Bible - ASV, n / d).

In the Catholic and Protestant traditions, death is seen as God's punishment for original sin.

We also note this personification of death: *"Finally, death itself came, roaring like a lion and very terrible in appearance; she looked like a man, only she had no body and was made up of bare human bones only. With her were various tools for torment: swords, spears, arrows, scythes, saws, axes, and other tools unknown to me. My poor soul trembled when it saw this. The holy angels said to death: "What are you delaying, free this soul from the body, free it quietly and soon, because there are not many sins behind it." In obedience to this order, death approached me, took a small cord, and first of all cut off my legs. My arms, then gradually cut off my other members with other tools, separating composition from composition, and my whole body became dead. Then, taking an adze, she cut off my head, and it became as if a stranger to me, for I could not turn it around. After that, death made some drink in the cup and, bringing it to my lips, forced me to drink. This drink was so bitter that my soul could not bear it - it shuddered and jumped out of the body as if forcibly torn out of it".* (The Ordeal of St. Theodora, n / d).

In turn, there is also the reverse process - repentance (metanoia, change of mind) - "revival of the soul" when it turns to "Life", that is, God, performed by the synergy of the will of man and the Spirit of God in the sacrament of Communion of the Holy Gifts, which is appropriate to compare with convolution scenario with increasing potential (Chapter 1)[3].

We summarize the data obtained in Table 5.

Table 5. Comparison of biblical ideas with the soliton-wave model of a person

Human components	Compliance with the soliton-wave model
Soul	The wave part of the system
Body	The solution part of the system

continued on following page

Table 5. Continued

Human components	Compliance with the soliton-wave model
Spirit	Intuition - Level 1 Channel
	Ego is a 2nd level channel
	Reason and will - channels of the 3rd level
Soul	Nous is a channel of the 4th level
	Feelings, emotions - channels 5, 6 levels
Death is a mystery	The transition from the dominance of a higher level of consciousness to a lower one within the framework of the sweep scenario with the loss of potential, appearance, and acceleration of time
	Some creature fixing such a transition
Repentance is also the "Second Baptism" and Communion	The transition from the dominance of the lower level of consciousness to the higher one in the framework of the convolution scenario with an increase in potential with no time at higher levels

Thus, the biblical understanding of human nature is surprisingly close to the soliton-wave model. The idea of multiple deaths as transitions from one stable state of consciousness to another will be explored in Chapters 5,6.

THE TIBETAN BOOK OF THE DEAD ANTHROPOLOGICAL IDEAS

The Tibetan Book reflects the Buddhist notions that developed in an era described as a "weakening of intuition, strong egos, will and reason, and a growing mind." Geographically, it covered both Hindustan, China, and the ancient Oikumene; in content, it included Taoism, six schools of Indian philosophy, and ancient natural philosophy. Therefore, it must be remembered that, with its specifics, the Tibetan Book of the Dead represents a very ancient, broad, and deep tradition. It, and not just the text of the Book, will be considered.

In the Western world, these ideas did not enter the modern scientific paradigm and were preserved (with various distortions) only in hermetic communities and the history of philosophy; to the East, they are an operating paradigm with "operational" implications for practice.

By the tradition in its Hindustani version, the Tibetan Book of the Dead operates with different koshas (receptacles), which are associated with the five elements that determine the three doshas (body constitutions - in the medical sense of the word), are grouped into three shariras (body levels), including the forms bhuta (gross) and sukshma (subtle), constituting a single personality (Atma, Purusha).

These koshas, doshas, shariras, sukshmas, bhutas, atma, and purusha are in complex mutual relations with each other, with the chakras, gunas, and the Universe, which includes not only the physical worlds but also the diverse spiritual worlds inhabited by gods, demigods, demons and other living and dead creatures - forming a truly vast and non-linear open system!

It follows that any presentation or criticism of the entire array of representations will be incomplete and deliberately simplified (Mishra, 2019; Satpathy, 2018).

Now we can proceed to characterize the five koshas and some of their connections, as they are understood by modern authors. According to (Hinduism. Anthropology, n / d): *"Hindu anthropology presents a human as a complex structure of the interaction of spiritual and material principles. The true essence of human - Atman is identical to the universal divine principle - Brahman. However, avidya (ignorance) hides this fact from human consciousness and binds it to the physical body, emotions, and conceptual thinking. According to Vedanta, the spiritual essence of a person is hidden by five koshas. Each kosha gives a person the opportunity to function and develop on different planes of being"* (Figure 7).

Figure 7. Human as a complex structure according to (Hinduism. Anthropology, n/d)

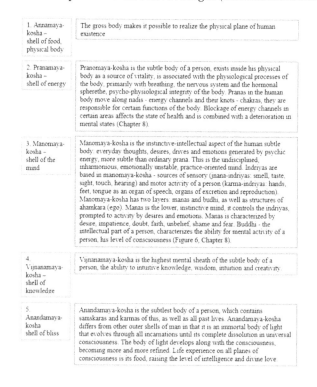

The essence or core of the soul is Atman, which is the innermost and immutable being of man - pure consciousness. The Atman has never been created, it does not change or evolve. He is eternally identical to the universal divine consciousness (Brahman). Anandamaya-kosha by its nature is only an emanation of the divine principle, which is destined to return to it again. This is the main goal of the development of the individual soul - differential self-consciousness by the Absolute of oneself. The question of the need for this process does not make sense - it's just a lila - a divine game (Hinduism. Anthropology, n / d).

Koshas are combined into three bodies for the soul, called sharira (Figure 8).

Figure 8. Three shells for the soul according to (Hinduism. Anthropology, n / d)

1. Sthula-sharira – physical body	Annamaya-kosha	The sthula sharira represents the physical body that man manipulates in the material world
2. Sukshma sharira, or linga sharira – the subtle body	Pranamaya, manomaya, and vijnanamaya koshas	Sukshma-sharira, the non-physical body that the soul puts on in order to function in the antarloka (subtle world). The pranamaya, manomaya and vijnanamaya koshas enter the subtle body during the period when the soul is physically incarnated. After death, the body of two shells, manomaya and vijnanamaya, continues to exist, since the pranas disintegrate simultaneously with the death of the physical body.
3. Karana-sharira – causal body	Anandamaya-kosha	Other shells dissolve into the anandamaya kosha before a new rebirth or when a transition to higher planes of evolution takes place. The Karana-Sharira persists through the process of reincarnation and is itself the cause of the next incarnations. The cycle of reincarnation begins in the antarloka, where souls reside before birth and where they return after death. The number and nature of births depend on past deeds (karmas). At birth, the soul forgets all memories of past lives and subtle worlds, where it stayed between births. The cycle of reincarnation ends when all karma is dissolved and self-realization is achieved. Liberation from rebirth is called moksha.

Let`s continue citation (Hinduism. Anthropology, n / d):

Hindu tantras are devoted to the psychoenergetic dimension of a person: the system of energy channels (nadis), chakras, body pranas, etc. Nadis are both a material substratum (nerve fibres, blood, and lymphatic vessels) and channels of vital energy not connected with a material carrier. Chakras are the intersection of energy channels, associated with acupuncture points, as well as the joints of the physical body. Large chakras - Muladhara (base of the spine); Svadhisthana (just below the navel); Manipure (solar plexus); Anahate (heart center); Vishuddhe (throat); Ajna (the point between the eyebrows); Sahasrara (top of the head) correspond to various nerve plexuses and endocrine glands. The existence of 72,000 nadis is recognized, among which there are three main channels - Pingala (sympathetic nervous system), Ida (parasympathetic nervous system), and Sushumna. Sushumna is the main nerve current that runs along the spine from its base (Muladhara chakra) to the crown of the head (Sahasrara). This channel is activated only when Ida and Pingala are balanced, which is achieved through the practice of yoga. The tantras describe specific practices aimed at raising the Kundalini (energy that is in a latent state) through the Sushumna to the Sahasrara chakra, which transforms all spheres of consciousness: spiritual, physical, emotional, and material.

Ayurveda Shastra explains how the five main pranas govern the physiological activity of all systems and organs. The main life force is localized in the heart and is called prana-vayu (governs the process of breathing, the work of the heart, and blood circulation in large and medium-sized blood vessels). Apana-vayu (lower abdomen) inspires and controls the organs of excretion and reproduction. Samana Vayu (the navel region) is responsible for digestive activity. Udana-vayu (throat) helps to exhale, and pronounce sounds and controls the nutrition of the brain, the development of memory, and mental abilities. Vyana vayu (distributed throughout the body) ensures the flow of blood and other fluids to all parts of the body and controls capillary circulation and perspiration.

Let us correlate the data with the soliton-wave coordinate system (Table 6).

Table 6. Comparison of "natural-philosophical" ideas with the soliton-wave model of a person

Concepts	Shariras, Koshas, Chakras	Correspondence to the soliton-wave model
Spirit	Karana sharira	The wave part of the system, levels 2-3
Soul	Sukshma sharira	The wave part of the system, levels 4-6
Body	Sthula sharira	The solution part of the system
The physical body, systems, organs, earth elements	Annamaya kosha, muladhara	The solution part of the system
Biofield, vitality, emotions, element of water	Pranamaya kosha, svadhisthana	Wave part of the system, levels 5-6
Practical mind (nous), instinctive desires, feelings, element of fire	Manomaya kosha, manipura	Wave part of the system, level 4
Will, reason, motives	Vijnanamaya kosha, anahata	Wave part of the system, level 3
Ego, joy of existence	Anandamaya kosha, ajna	Wave part of the system, level 2, actual existence
Intuition, unity	Atma, Sahasrara	Wave part of the system, level 1, the potential existence

We state a rather natural correlation with channels and levels of the soliton-wave model.

Now let's make a pivot table that allows you to compare source data with each other. At the same time: representations of the Paleolithic, Neolithic, Popol Vuh, and fairy tales are not shown as uninformative, as well as calibrations and correlations with the soliton and wave components discussed above. Table 7 shows only the relationship between the levels of the subjective space concerning the soliton-wave coordinate system.

Table 7. Comparison of ideas about the structure of a person in the considered sources

Levels	Amulets of Inanna	Shells of Egypt	Spirit, soul, body		Atma, Shariras, and Koshas	
1	Crown of Eden	Ren	Spirit	Intuition	Atma	
2	Signs of Dominion and Judgment	Ib		Ego	Karana sharira	Anandamaya kosha
3	Necklace			Reason and will	Sukshma sharira	Vijnanamaya kosha
4	Pendant on the chest		Soul	Nous		Manomaya kosha
5-6	Grid "To me, man, to me"	Ka		Feelings, emotions		Pranamaya kosha
Body[4]	Body	Khet		Body	Sthula sharira	Annamaya kosha

As can be seen from Table 7, the representations of different eras are like each other.

FUTURE RESEARCH DIRECTIONS

Using the same sources and method, it is necessary to reconstruct post-mortem scenarios and compare them with each other and with the generalized post-mortem scenario substantiated in Chapter 2. These tasks will be solved in Chapters 5, 6.

CONCLUSION

The authors feared that the research method described in Chapter 12 could show the dissimilarity of ideas from different eras. Testing the post-mortem scenario described in Chapter 11 would then be impossible. However, despite the external dissimilarity (calibrations), these ideas demonstrated internal compatibility (invariants). The HEM used as a "coordinate system" has created a convenient basis for comparing these concepts and the soliton-wave model. It is surprising how insignificant the influence of the evolution of human consciousness (Kovalyov et al., 2020), as far as national, religious, and cultural differences, was on ideas about post-mortem existence. Now back to the conclusions:

- The applicability of the method of verification of post-mortem scenarios using universal human experience to determine the structures that continue to exist after death is shown.
- Structures that continue posthumous existence were identified and compared with each other and the HEM (Tables 1-6).
- In various sources structures - invariants - practically do not change, but their descriptions - calibrations - change.

These claims will be tested in Chapters 5, and 6, specifically devoted to the events of the post-mortem scenario analysis of the sources.

We now return to the problem of the sufficiency of evidence. It would be correct to formulate it this way: is the revealed similarity of structures a sufficient reason to consider the potential possibility of the existence of a post-mortem scenario outlined in Chapter 2 as actual?

The similarity of objects in science can be interpreted in different ways. Consider an example.

Signs were found on the rocks of the Stone Grave; experts recognize them as similar to Mesopotamian cuneiform; even more exactly, they are identified as Elamite cuneiform (Figure. 9). The fact of their presence and similarity takes place. But what does this similarity mean?

Figure 9. The signs of Stone Tomb and Mesopotamian and Elamitian cuneiform. National Historical and Archaeological Reserve "Stone Grave". Ukraine, Melitopol

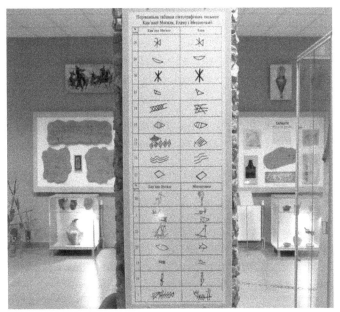

(Y.N. Kovalyov, with the permission of the museum staff)

Are the Elamites' ancestors locals in the Northern Black Sea region and then migrated to the Persian Gulf through the Caucasus (or the Balkans), Asia Minor, and all of Mesopotamia in prehistoric times? Or did the Elamite priests leave a record after a sacrifice brought in the "world-famous" sanctuary of the cult of the dead? Or was it visiting Elamite merchants - how do modern tourists visit the Sagrada Familia in Barcelona? Or maybe cuneiform signs are similar - how similar are spears and swords among all peoples and in all eras? Or are these manifestations of some "common archetype"?

Similarly, it is possible to evaluate the revealed sameness of post-mortem existing structures of subjective space in different ways. The similarity of ideas from different eras shows their belonging to a ternary connective leading to intuitive knowledge; they can also be considered as an expression of intuitive knowledge by K1 means. On the other hand, the revealed similarity may indicate the existence of a very deep and ancient gestalt, through which ideas about the afterlife are formed, and its close connection with the convolution scenario of a complex system. Other interpretations are also possible. Therefore, we cannot consider the revealed similarity as convincing proof for all of the post-mortem scenario's actual existence.

REFERENCES

Amin, O. S. M. FRCP(Glasg) (2016). Ashurnasirpal II performs religious rituals before the sacred tree. [Illustration]. *File: Ashurnasirpal II performs religious rituals before the sacred tree. From Nimrud, Iraq. 865-860 BCE. British Museum.jpg*. (n.d.). Wikimedia.org. Retrieved August 15, 2024, from https://commons.wikimedia.org/wiki/File:Ashurnasirpal_II_performs_religious_rituals_before_the_sacred_tree._From_Nimrud,_Iraq._865-860_BCE._British_Museum.jpg

Coleman, G. (Ed.). Jinpa, T. (Ed.), Dorje, G. (Transl.), & Dalai Lama (Comment.). (2007). *The Tibetan Book of the Dead: First Complete Translation.* Penguin Classics

Collins, A. (2014). *Göbekli Tepe: The Origin of the Gods.* Bear & Company.

Dasgupta, S. A. (2014). *History of Indian Philosophy.* (5 vol.). Cambridge University Press. (Original work published 1922)

Dedović, B. (2020). *"Inanna's Descent to the Netherworld": A centennial survey of scholarship, artifacts, and translations.* Digital Repository at the University of Maryland. https://doi.org/DOI: 10.13016/ur74-yqly

Foster, B. R. (2001). *The Epic of Gilgamesh.* W.W. Norton & Company.

Freud, S. (1978). *The Ego and the Id and Other Works (1923–26).* v. XIX. In The Standard Edition of the Complete Psychological Works of Sigmund Freud. London the Hogarth Press. (Original work published 1923)

George, A. (2014). *Epic of Gilgamesh.* Penguin Classics.

Goelet, O. (Transl.), Faulkner, R. (Transl.), Andrews, C. (Pref.), Gunther, J. D. (Intro.), & Wasserman, J. (Foreword). (2015). *Egyptian Book of the Dead: The Book of Going Forth by Day: The Complete Papyrus of Ani.* Chronicle Books.

Heroes of Sumerian-Akkadian myths. [Illustration]. Used files:

4–. Enkidu. Author: Amin, O. S. M. FRCP(Glasg). (N.d.). Wikimedia.org. Retrieved August 15, 2024, from https://commons.wikimedia.org/wiki/File:Enkidu,_Gilgamesh%27s_friend._From_Ur,_Iraq,_2027-1763_BCE._Iraq_Museum.jpg/*Holy Bible - American Standard Version.* (2019, April 1). Holy Bible - American Standard Version - ASV. https://holy-bible.online/asv.php? Jung C.-G. (2002). *Psychological Commentary on The Tibetan Book of the Dead.* In *Jung on Death and Immortality.* Princeton University Press. (Original work published 1923). https://doi.org/DOI: 10.1515/9780691215990-004

2–. Ereshkigal. Photographer: BabelStone. *File: British museum queen of the night. Jpg*. (n.d.). Wikimedia.org. Retrieved August 15, 2024, from https://commons.wikimedia.org/wiki/File:British_Museum_Queen_of_the_Night.jpg

3–. Gilgamesh. Photographer: U0045269. Wikipedia contributors. (n.d.). *File:O.1054 color.jpg*. Wikipedia, The Free Encyclopedia. https://en.wikipedia.org/wiki/File:O.1054_color.jpg

1–. Inanna. Author: Sailko. *File: Ishtar on an akkadian seal.Jpg*. (n.d.). Wikimedia.org. Retrieved August 15, 2024, from https://commons.wikimedia.org/wiki/File:Ishtar_on_an_Akkadian_seal.jpg

Keightley, D. N. (2004). *The Making of the Ancestors: Late Shang Religion and Its Legacy*. In *Religion and Chinese Society* (Lagerwey, J. Edit.). (2 v.). The Chinese UP DOI: 10.2307/j.ctv1z7kkfn.4

Kovalyov, Y., Mkhitaryan, N., & Nitsyn, A. (2020). *Self-organization of the human mind and the transition from paleolithic to behavioral modernity*. IGI Global International Publisher. DOI: 10.4018/978-1-7998-1706-2

Leroi-Gourhan, A. (1967). *Treasures of Prehistoric*. Harry N. Abrams.

Leroi-Gourhan, A. (2015). *Les religions de la préhistoire* (7th ed.). PUF.

Levi-Strauss, C. (1966). *The Savage Mind*. University of Chicago Press.

Lewis-William, D. (2002). *The Mind in the Cave*. Thames and Hudson.

Mishra, Y. (2019). Critical Analysis of Panchakosha Theory of Yoga Philosophy. *World Journal of Pharmaceutical Research*, 8(13), 413. DOI: 10.20959/wjpr201913-16152

Radhakrishnan, S. (2009). *Indian Philosophy*. (2 vol.). Oxford University Press. (Original work published 1923)

Satpathy, B.Dr. Biswajit Satpathy. (2018). Pancha Kosha Theory of Personality. *International Journal of Indian Psychology*, 6(2). Advance online publication. DOI: 10.25215/0602.105

Schmidt, K. (2020). *Sie bauten die ersten Tempel: Das rätselhafte Heiligtum am Göbekli Tepe*. C.H. Beck.

Thurman, R. (2011). *The Tibetan Book of the Dead: Liberation Through Understanding in the Between*. Bantam.

ADDITIONAL READING:

Albedil, M. F. (1991). *Zabytaja cyvylyzacyja v dolyne Ynda* [A Forgotten Civilization in the Indus Valley]. Nauka.

Bader, O. N. (1967). *Pogrebeniya v verhnem paleolite i mogila na stoyanke Sungir.* [Burials in the Upper Paleolithic and a grave at the Sungir site]. *Soviet Archeology 1967*(3). https://arheologija.ru/bader-pogrebeniya-v-verhnem-paleolite-i-mogila-na-stoyanke-sungir/

Danylenko, V. M. (1986). *Kamyana Mohyla* [Stone tomb]. Naukova Dumka.

Dergachev, V. A., & Manzura, I. V. (1991). *Poghrebaljnye kompleksy pozdnegho Trypoljja* [Burial complexes of late Trypillia]. Shtiintsa.

Dyakonov, I. M. (2006). *Epos o Ghyljghameshe (O vse povydavshem)* [The Epic of Gilgamesh ("About who has seen everything")]. Nauka.

Granin, R. S. (2019). Eskhatologhycheskaja paradyghma y ee struktura. [Eschatological paradigm and its structure]. *Interdisciplinary Research*, 2019, 133–147.

Hinduism. Anthropology. (n/d). Retrieved August 15, 2024, from https://mybiblioteka.su/6-144473.html

Human triparties. (n/d). Retrieved August 15, 2024, from https://azbyka.ru/otechnik/prochee/dusha-chelovecheskaja-polozhitelnoe-uchenie-pravoslavnoj-tserkvi-i-svjatyh-ottsov/19

Khazarzar, R. Library. (n / d). Retrieved August 15, 2024, from http://khazarzar.skeptik.net/books/shumer/inanna.htm

Kindness (1992). (5 vol.). Holy Trinity Sergius Lavra.

Korvin-Piotrovsky, A. G. (2008). *Trypiljsjka kuljtura na terytoriji Ukrajiny* [Trypillia culture on the territory of Ukraine]. Institute of Archeology of the National Academy of Sciences of Ukraine.

St. Luka (Voyno-Yasenetsky). (2013). *Duh, dusha i telo.* [Spirit, soul, and body]. Dar.

Mykhailov, B. D. (2005). *Petroghlify Kam'janoji Moghyly: Semantyka. Khronologhija. Interpretacija* [Petroglyphs of Kamyana Mohyla: Semantics. Chronology. Interpretation]. MAUP.

Mytarstva sv, Teodory. [Ordeals of St. Theodora]. (n / d). Retrieved August 15, 2024, from https://lib.pravmir.ru/library/readbook/1959

Shaposhnikov, A. K. (Ed.). (2003). Drevneeghypetskaja Knygha Mertvykh. Slovo ustremlennogho k Svetu. The Ancient Egyptian Book of the Dead. The word of one striving towards the Light. Eksmo Publishing LLC

St. Gregory Nyssky. (1862). *Tochnaja ynterpretacyja Pesny Solomona*. [The exact interpretation of the Song of Solomon]. In *Creations of the Holy Fathers in Russian translation*. v. 39: Works of St. Gregory of Nyssa. part 3, 298

St. Gregory Palamas. (2004). Vsechestnoj v monakhyne Ksenyy, o strastjakh y dobrodeteljakh y o plodakh umnogo delania. [To the all-honorable in the nun's Xenia, about passions and virtues and fruits of smart doing] In *Philokalia* (Vol. 5). Sretensky Monastery Publishing House.

St. Gregory Palamas. (2011). *Tryady v zashhytu svjashhennbezmolvstvuyuschyh*. [Triads in defense of the sacred silent ones]. Academic project.

St. Theophan the Recluse. (n / d). *O duhe i dushe*. [About spirit and soul]. Source: https:/ /nasledie77.wordpress.com/2013/11/04/theophan-recluse-about-spirit-and-soul/, n / d

Tibetskaya kniga mertvyh. [Tibetan Book of the Dead]. (1992). Chernyshev Publishing House.

Vasiliadis, N. (2012). *Taynstvo smerti*. [Sacrament of death]. Publishing House of the Holy Trinity Sergius Lavra. (n / d). Retrieved August 15, 2024, from https://pravoslavnoe.uaprom.net/p782708550-vasiliadis-tainstvo-smerti.html

Zubov, A. (2017). *Doystorycheskye y neystorycheskye relyghyy* [Prehistoric and non-historical religions]. RIPOL Classic.

ENDNOTES

[1] The cycle of Indian myths about Purusha correlates different parts of his body with the places of the Universe, gods, elements, chakras, and varnas (Chapter 5). This makes interpretation easier. How legal is its use? There are two arguments "for" - the culture of ancient people is much more homogeneous than it is now, and many symbols are similar; there are assumptions about the connections of the Sumerians with the Indians, therefore, borrowing is not excluded. The authors find it difficult to assess the weight of these arguments.

[2] This description resembles one of the ideas of Taoist alchemy.

[3] Compare with Egyptian ideas: after death, Osiris endows the deceased with life and body (Ah and Sah), making him able to stay with the gods; a Christian, while still living, takes communion with the Flesh and Blood of God, gaining a body and life for eternal staying with Jesus Christ.

[4] Body levels duplicate levels of subjective space (Chapter 8); they are not shown

Chapter 5
Events Checking:
What Old Artifacts and Middle Eastern Texts Can Tell

ABSTRACT

Verification of the invariants – events, and transitions – predicted within the general and particular post-mortem scenarios (Chapter 2) showed their correspondence to data from sources of different cultures and eras. At the same time, the difference in particular scenarios is associated with changes in human consciousness from era to era, which changes the initial data for the convolution process. Calibrations allow us to find out how consciousness perceives the invariants of the general post-mortem scenario.

BACKGROUND

Sources were used by their ranks, supplementing them as needed. We will pay more attention to data about the post-mortem events.

From this viewpoint, the surviving artifacts of the pre-literate period, that is, the Paleolithic, Mesolithic, Neolithic, and partly Chalcolithic, are of interest (Bader, 1967; Collins, 2014; Danylenko, 1986; Mykhailov, 2005; Schmidt, 2011; Old European culture, 2016; Johnston, 2006). However, the reconstructions made on their basis and with the involvement of ethnographic material are usually debatable.

The first described and deciphered "topographies" of the underworld and the fate of the souls of the dead begin with the Sumerian-Akkadian myths (Foster, 2001; Dyakonov, 2006; Dedović, 2020).

More information is contained in the complex of ancient Egyptian sources (Wallis Budge, 11967/1895; Mercer, 1952; Shaposhnikov, 2003; Granin, 2019; Goelet et al., 2015), and here modern reconstructions are based on rich factual material,

DOI: 10.4018/979-8-3693-9364-2.ch005

including, in addition to texts, images, tombs, numerous ritual objects, mummies and posthumous gifts are sometimes treasures, as in the tomb of Tutankhamen.

Further, these are the monotheistic religion's Holy Scriptures data, their subsequent exegesis, liturgical and prayer practices and rituals, the Holy Father`s texts, and their modern comments (Holy Bible (ASV); Vasiliadis, 2012; St. John Chrysostom, 2000; Ig. Hilarion (Alfeev), 2001; Prayers for the dead, 2016; St. Macarius of Alexandria, 1831; Ordeals of St. Theodora, n / d).

All sources will be used to verify post-mortem scenarios, including invariants and calibrations; the sources will be compared with post-mortem scenarios (Chapter 2) and each other.

ARTIFACTS OF THE PRE-HISTORIC PERIOD AND POSSIBLE SCENARIOS

The activity of the channels of human interaction with the environment has changed throughout its evolution; at first, the intuition channel was the most active (Kovalyov et al., 2020).

Based on this, it can be assumed that a person of that time did not see a particular problem in death, feeling the continuous unity of the genus, as if growing through the worlds of the living and the dead (one of the interpretations of the World Tree). The post-mortem scenario could be described in one sentence: as leaves fall in autumn and grow in spring, as the sun sets in the evening and rises in the morning, a human dies and is reborn again. This follows from the relatively simple burials and their connection with the fertility cult of Mother Earth in its pre-agrarian form. There were other associations - death was compared with sleep (even Jesus speaks of the deceased Lazarus as having fallen asleep), as well as near-death experiences.

Such representations are quite consistent with reincarnation scenario 2 (Chapter 8).

Gradually, however, intuition weakened (but remained much stronger than in our days) and the channels of ego, will, and reason intensified. This led to the emergence of symbolic activity; human consciousness was highly associative and "magical". For a modern researcher, this creates considerable cognitive problems: if we see just a seagull in a seagull, then an ancient person saw a spirit in it (named Jonathan Livingston with all the background), recognized the first ancestor of the genus, understood its flight as an omen, etc. Using the terminology defined in Chapter 12, ancient humans saw not a sign but a symbol, not a "point" but a "direction", not a projection but a ternary connective, a multiple-gauge invariant, not Backus's normal form.

Ideas about the afterlife changed with a gradual change in the consciousness state. Here is a description of Upper Paleolithic burial at the Sungir site (Bader, 1967): *"The grave was dug, counting from the upper ocher platform to a depth of 60–65 cm, cutting through about 15 cm of the soil and cultural layer and 48–50 cm of the underlying yellow sandy loam. It had the shape of a strongly elongated oval, the maximum dimensions near the bottom were 2.05 x 0.70 m, and was oriented from the southwest to the northeast. The headboard was 8-9 cm higher than the bottom of the grave under the feet. The skeleton of a 55-65-year-old man lay in the grave in an extended position on his back, with his head to the northeast, with his hands crossed on the pelvic bones, densely covered with red ocher powder and with an exceptionally rich inventory of bone and stone items (Figure 1).*

Figure 1. Burial at the Sungir site

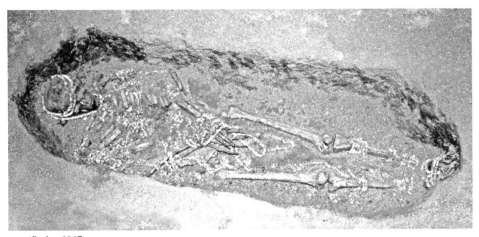

(Bader, 1967)

After the grave was dug, its bottom was sprinkled with coal and ash. Then the grave was densely sprinkled with red ocher powder, which in some places formed a layer up to 3 cm thick and partly covered the lower part of the grave walls. Only after that the deceased was laid in the grave and again heavily sprinkled with ocher. In addition to numerous decorations on clothes and bracelets on his hands, a re-touched flint knife was placed between the legs of the corpse; he, like the skeleton, lay on a layer of ocher and was covered with it from above. On the second layer of ocher, which covered the skeleton and the bulk of the jewelry, lay several more rows of mostly larger beads; they can be associated with some kind of cover, probably clothing, which was not put on the corpse, but covered it from above. After that, the other was sprinkled for the third time. After filling the grave, during which at least three more times ocher was thickly sprinkled, the surface of the grave, apparently at the level of the soil surface of that time, was again and most densely covered with ocher, which marked the place of the grave in the parking lot, and a stone and human skull. It belonged to a Caucasian woman. Mention has already been made of the ritual burial of skulls separated from the body custom, that arose in the Upper Paleolithic (Placard and Arcy-sur-Cure, France); obviously, it was assumed that a person's soul lives in it".

Anthropological finds at the Sungir site are represented, in addition to the mentioned skull and skeleton, by five more skeletons, a femur, a fragment of a femur, and as rich grave goods - about 80 thousand ornaments were found. The paired burial of adolescents is the most interesting (Figure 2).

Figure 2. Paired burial of adolescents at the Sungir site: burial (bottom - reconstruction of outer clothing, taking into account the location of the preserved bone stripes)

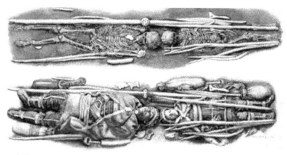

(Primitive Sungir, n / d)

According to anthropometric characteristics, the Sungir people were Cro-Magnons, but the problem of the exact age of dating the burials has not been finally resolved; radiocarbon dating of ^{14}C repeatedly performed in different laboratories gave a spread from 23830 ± 220 to 30000 ± 550 years ago (Kuzmin, n / d).

Here you can trace the development of ancient ideas. "The path of the earth" (in later myths - the "path of the moon") and the "path of the sun" are still combined (burial in the ground, sprinkling with ocher, the orientation of the corpse), but have their attributes. The soul is separated from the body and a certain material object (skulls, churingas) is needed for its manifestation in the world of the living; on the contrary, to project the properties of material objects in the world of the dead, their burial is necessary (costumes, jewelry (amulets?), weapons and other inventory). Relatives take care of the corpse as a member of the family, but get rid of it (fear of the world becoming alien); they also try to preserve the possibility of communication (specially preserved skulls).

Something fundamentally new also appears: judging by the richness and composition of the grave goods, it is assumed that the deceased "leaves for a long time" - or forever. This already takes the existence of the deceased beyond the reincarnation scenario, which significantly increases the importance of knowledge about the structure, events, and transitions of the soul in the world of the dead. Let's view how space is shown in Paleolithic paintings (Figure 3).

It is not like the usual three-dimensional space, differing from it in the following features (Kovalyov et al. 2020):

- Immediacy, the brightness of colors, the liveliness of plots, and the space itself - dynamic, polymorphic, chaotic in its uncertainty.
- Two-dimensionality and polymorphism. There are no elements of perspective images (point of view, skyline, vanishing points, main beam). But an illusion of depth is created - due to the relief, texture, and shades of colors.
- Dynamics, lack of any symmetry or center.
- There are no points of reference that would allow us to fix the sky or the earth, the movement of time, or a point of view. Let's make a reservation - the top and bottom are quite distinguishable for some scenes.
- Distortion of natural scales and proportions: figures seem to grow out of walls and relief.
- A deep understanding of the "soul" of animals, and unity with them. Although totemism and the domestication of animals usually date back to a later time, it can be assumed that such unity was a necessary condition for domesticating an animal or feeling it as an ancestor of a kind.
- The presence of polymorphic images.
- Abstract images.
- Magical or cult scenes.

Figure 3. Free placement of drawings and use of the relief of the cave walls. Lascaux cave, France.

(Lascaux cave, n / d)

Such features not only illustrate "magical thinking" but show what the world of the dead looked like in the eyes of Paleolithic artists - chaotic, dynamic, timeless, but filled with a peculiar life. This is evidenced by the location of many images in caves associated with the womb of Mother Earth.

The funeral rite from the time of the Upper Paleolithic should likely be considered a magical action designed to provide the deceased with an existence like the earthly one in the afterlife while maintaining the lifetime status and receiving "feedback" from him as some benefits for the family. Sometimes the Paleolithic painting depicting animals is interpreted as an "eternal sacrifice" to the spirits of the ancestors for this very purpose. So, the ideas of the Upper Paleolithic - modus vivendi of the spirit of the deceased with the realities of the afterlife with magical support from the world of the living and its existence there for a long time can already be compared with the post-mortem scenario of stabilization - the existence of a wave component, in the presence of exchanges and interactions, within the framework of the realities of the "wave world", after the soliton-wave system destruction.

In the subsequent period, the growing alienation from the world of the dead (weakening of intuition), the growing fear of death (growth of the ego), the separation of the "earthly" ("lunar") and "solar" paths with their binary assessment (strengthening of the mind) become more and more noticeable. Ordinary people went along the first path, and only outstanding personalities with a strong will could go along the second - rulers, shamans, magicians, priests, heroes, messengers of spirits, or gods. They "organized the movement", using funerary structures, inventory, and rituals as magical tools. This is expressed as follows.

Firstly. it was necessary to "aim" at the right place (and time). It can be assumed that this explains the shape of the Neolithic burial structures - dolmens (Figure 4). Other interpretations - the "house of the soul", and the womb of Mother Earth - are compatible with this assumption.

Figure 4. Different types of dolmens

(Ivan Chelovekov, n / d)

Likely, the astronomical references to Stonehenge and several other objects are due to the desire to ensure that the chosen dead get on the "sunny path". In any case, it is known that the territory of Stonehenge (Figure 5) was used for burials even before the construction of the famous complex (from about 3030-2880 BC), and the last burials date back to 2570-2340 BC. e.; in total there were approximately 240 burials, possibly descendants of one family (The Stonehenge Mystery, 2008).

Figure 5. Neolithic Stonehenge complex (before restorations). View of the ruins in a photograph taken in July 1877. United Kingdom

(Smyrnov, n.d.)

Further, it was necessary to take certain actions, the nature of which can be judged by the burial complex of Brou-on-Boine, in particular, Newgrange - a mound with an inner corridor and a central chamber (Figure 6), known for the fact that on the day of the winter solstice, the sun's rays fall exactly into the central chamber (3 and 4 in Figure 6).

There are comparisons of the complex with the female womb, which seem quite plausible when looking at diagram 3 in Figure 6, and Figure 4, Chapter 7, (Old European Culture, 2016).

A man's burial was also found[1], and DNA analysis gave important results for understanding the functions of the structure: *"DNA from a middle-aged man buried in 3200 B.C.E. at the center of this mighty mound suggests otherwise. His genes indicate he had parents so closely related they must have been brother and sister or parent and child"* (Curry, 2020).

From the totality of the given data, the three-spiral ornament (Figure 6.6), as well as from the text of many myths, it is clear that with the help of the Newgrange structures, a ternary connection was provided between the worlds of Heaven, Earth, and the Underground, the living, the dead and spirits (or the god of the Sun), while the unity of the worlds was considered as an act of fertilization. This is a first approximation, and now let's try to clarify the details.

The mound itself is a symbol, that is, a boundary space. As a boundary space, it contains a few projections. So, the projections of the world of living and Mother Earth are a mound - the womb, a corridor, and a chamber - the reproductive system. The image of the underworld ("the earthly path", the world of the dead) was projected as a burial or sacrifice. The image of the Sky, or specifically the Sun, was projected as a sunray. Penetration of the beam through the corridor into the chamber - the implementation of the connection of the worlds in the act of fertilization.

The priest (subject), using this toolkit and introducing various calibrations (rituals, spells), had the opportunity to influence the reality of all three worlds (object), organizing and inverse effects on himself and in the world of the living.

Figure 6. Newgrange Tomb, Newgrange, Ireland. Approximately 3200 years. BC. Ireland: [Used files in references]. 1 - Modern view after restoration. (Newgrange, 2018); 2 – The design of the mound, corridor, and chamber (Source: Newgrange: fairy mound in Ireland, n / d); 3 – The path of the sun's rays on the day of the winter solstice (Rick Doble, 2015); 4 - Sun rays in the corridor on the day of the winter solstice (Newgrange Winter Solstice, 2012); 5 - View of the central chamber. (Newgrange, 2018); 6 – A stone with images of spirals at the entrance (Rick Doble, 2015)

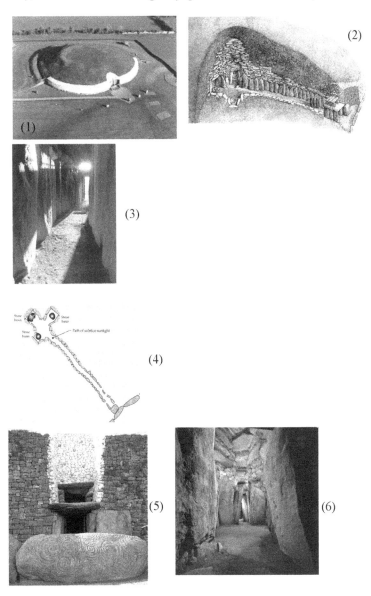

For example, the well-being of the agrarian inhabitants depended on the fertility of the land - and the very fact of the sacred fertilization of the Earth by the Sun on the day of the winter solstice was supposed to ensure (renew) such fertility. Or the dead ruler could be sent to the "solar path", having received a new, solar body in the act of sacred fertilization (he died for the living and came to life for the dead). Or a certain messenger could be sent to Heaven (or the Sun) along the "solar path" with emergency request - in this case, a human sacrifice took place. Also, a ritual could be performed to ensure the legitimacy of power. It supposedly looked like this. On the day of the winter solstice, in a chamber, in the light of the sun (the phenomenon lasts from 14 to 21 minutes), the real fertilization of the "queen" by his brother (father, uncle) "king" could be carried out. Fertilization had immediately acquired a sacred character - in the optics of Neolithic consciousness, the born heir was rightfully the son of the Son of Heaven (Sun) and Earth, and the Lord of the three worlds - such a title would be envied by all historical and modern rulers, including the emperors of China and Tahuantinsuyu! Moreover, the presence of the first ancestor - the Sun - and the deceased ancestor-ruler (conception on the coffin) ensured his constant incarnation in the face of a conceived child, perhaps even "from the creation of the world"! There were probably other rituals as well.

Now let us ask the question: why was inbreeding fraught with poor heredity in the ruling family necessary at all? In the mind of Neolithic man, securing the connection of the worlds had the highest priority, and the ruling family itself was a kind of victim, in a strong theocratic society, perhaps not entirely voluntary. Using SWM and the data on reproduction and inheritance (Chapter 6), modern humans might suggest that the advantage lies in creating the conditions for the resonance necessary for fertilization - both sacred and real.

Remaining in working condition for 5000 years, a programmable magical machine is the pinnacle of the achievements of an era when intuition, ego, will and reason did not need books to create tools like Newgrange; it still waiting to be used - but the change human consciousness cannot "turn it on".

It was also possible to "get to the right place" with the help of a guide or metapomp (μεταπομπ) - an expert on the ways in the world of the dead. This guide originally was a shaman (Mykhailov, 2005). And this gives rise to the problem of getting, survival, and activity of a living being in the world of the dead, which deserves separate consideration. There are no obstacles to use for the study of SWM, the theory of self-organization, and the apparatus of the ternary connective.

Then, the conductor functions are passed to some "border creature", such as a dog or a jackal, or even a snake, leading a burrowing lifestyle. Later they are described as Kerberos, and Charon, carrying souls across the Styx, and so on. The guide has one more function - it not only takes the souls to the place of their posthumous stay but also punishes them (in the Inuit legend, beats them for about a year) or takes them to court. So, he turns from a border creature into an angel, dog-headed St. Christopher, or the god Anubis, as the prototype of St. Christopher.

Summing up, note that the considered facts and reconstructions fit into the re-incarnation 2 and stabilization scenarios, predicted in Chapter 11. In general, the paradigm of our perception of the world largely comes from the ideas of people of the Stone Age. Without digressing from the study, note the posthumous scenarios, events, and situations, passed into the myths and religions of the subsequent period. The most important include:

- Division of the path in the world of the dead into "the path of the sun" for the elect and "the path of the earth" for the rest".
- The influence (ritual or moral) of events in the world of the living on the state of the souls of the dead.
- Guide to the world of the dead.
- Living in the world of the dead.
- Finding a new body for life in the world of the dead.

There are details, that are repeated in religious texts: "exit to the light", "body of light" and others.

THE LIVING IN THE WORLD OF THE DEAD: INANNA AND GILGAMESH

The myths about Inanna and Gilgamesh are approximately synchronous with the construction of Newgrange and Stonehenge. These are interesting for several reasons: the main characters go to the kingdom of the dead each in their own way, the stages of passage are described, there are guides, and finally, both Inanna and Gilgamesh are alive.

Inanna's Path

Let's start with the earlier myth of Inanna, already known to Gilgamesh. So (Dedović, 2020):

[1-5] From the great heaven she set her mind on the great below. From the great heaven, the goddess set her mind on the great below. From the great heaven Inanna set her mind on the great below. My mistress abandoned heaven, abandoned earth, and descended to the underworld. Inanna abandoned heaven, abandoned earth, and descended to the underworld.

From this, she was going "the way of the earth."

Inanna conducts a preliminary preparation consisting of three stages:

1. First: *[14-16] She took the seven divine powers. She collected the divine powers and grasped them in her hand. With good divine powers, she went on her way.*

For a better understanding, let's draw a parallel with Jesus Christ. To defeat Hell and Death, he had to get their abode, for which he should become like a man: become a God-man; then renounce his Divine Immortality – *"And about the ninth hour Jesus cried with a loud voice, saying, Eli, Eli, lama sabachthani? that is, My God, my God, why hast thou forsaken me?"* (Matthew 27:46; Mark 15:34) - Jesus' dying words on the cross, quoted from Psalms 22:1; and only then die. Inanna does the same, but not for the salvation of mankind, but for becoming the mistress of two worlds.

2. Then Inanna puts on the magical charms described and compared in Chapter 4 with the levels and channels of a person's subjective space.
3. Finally, she instructs her servant to ensure the return from the realm of the dead if her strength proves insufficient:

[58-64] "In Eridug, when you have entered the house of Enki, lament before Enki: "Father Enki, don't let anyone kill your daughter in the underworld. Don't let your precious metal be alloyed there with the dirt of the underworld. Don't let your precious lapis lazuli be split there with the mason's stone. Don't let your boxwood be chopped up there with the carpenter's wood. Don't let young lady Inanna be killed in the underworld."[65-67] "Father Enki, the lord of great wisdom, knows about the life-giving plant and the life-giving water. He is the one who will restore me to life."

Pay attention to the appearance of certain "life-giving plants" and "the life-giving water " - artifacts necessary for the revival of myths of different peoples - can be interpreted as resuscitation.

The path itself follows:

[73-77] When Inanna arrived at the palace Ganzer, she pushed aggressively on the underworld gates. She shouted aggressively at the gate of the underworld: "Open up, doorman, open up. Open up, Neti, open up. I am all alone and I want to come in."

The gates of the underworld appear (biblical parallel: *"Lift up your heads, O ye gates; And be ye lifted up, ye everlasting doors: And the King of glory will come in"* (Psalms 24:7)), as well as a guide combining here the functions of the conductor and the guard.

Inanna tries to trick Neti:

[78-84] Neti, the chief doorman of the underworld, answered holy Inanna: "Who are you?" "I am Inanna going to the east." "If you are Inanna going to the east, why have you traveled to the land of no return? How did you set your heart on the road whose traveler never returns?"

[85-89] Holy Inanna answered him: "Because lord Gud-gal-ana, the husband of my elder sister holy Erec-ki-gala, has died; to have his funeral rites observed, she offers generous libations at his wake – that is the reason."

Seeing dead relatives is one of the typical near-death experiences (Chapter 10); there the experiences themselves were interpreted as pre-death experiences.

But the trick fails: warned by Neti, Erec-ki-gala unravels the plans of her rival sister and prevents them.

To open each of the seven gates, Inanna must pay with one of the amulets (Chapter 4), and on a date with Erec-ki-gala she comes disarmed and deprived of the will to fight: *"The afflicted woman was turned into a corpse. And the corpse was hung on a hook."*

The description is like a deep trance. There is an important circumstance: body preservation is a condition for revival. The soul and body seem to change places: the soul is eternal for the world of the living and cannot be destroyed there, but for the world of the dead the body is eternal and cannot be destroyed. But Neti, following the order of Erec-ki-gala, consistently "closes" the levels and channels of the subjective space. This explains the efforts of the ancient Egyptians (and not only them) to preserve the corpse; the revival of the dead before the Last Judgment, and so on - a full life is possible only within the framework of the emergent system of soul and body.

So, Inanna was defeated and fell into full power to Erec-ki-gala, and could not independently return from the world of the dead. But, in principle, she reached the very end of the stage of certainty 4, retaining the modality of the potential existence of her personality (Chapter 11). In addition, the body was also preserved. Therefore, the instruction given to her servant works. Tom manages to plead with Enki (all the other gods refuse, and Enki also agrees to help very reluctantly), and "resuscitation measures" begin. Enki himself does not dare to go to the underworld and creates two mysterious creatures for this, gives them a life-giving plant and life-giving water, and instructs in detail:

[226-235] Then Father Enki spoke out to the gala-tura and the kur-jara: One of you sprinkle the life-giving plant over her, and the other the life-giving water. Go and direct your steps to the underworld. Flit past the door like flies. Slip through the door pivots like phantoms. The mother who gave birth, Erec-ki-gala, on account of her children, is lying there. Her holy shoulders are not covered by a linen cloth.

Her breasts are not full like a cagan vessel. Her nails are like a pickaxe (?) upon her. The hair on her head is bunched up as if it were leeks.

[236-245] "When she says "Oh my heart", you are to say "You are troubled, our mistress, oh your heart". When she says "Oh my liver", you are to say "You are troubled, our mistress, oh your liver". (She will then ask:) "Who are you? Speaking to you from my heart to your heart, from my liver to your liver – if you are gods, let me talk with you; if you are mortals, may a destiny be decreed for you." Make her swear this by heaven and earth.1 line fragmentary

[246-253] "They will offer you a river full of water – don't accept it. They will offer you a field with its grain – don't accept it. But say to her: "Give us the corpse hanging on the hook." (She will answer:) "That is the corpse of your queen." Say to her: "Whether it is that of our king, whether it is that of our queen, give it to us." She will give you the corpse hanging on the hook. One of you sprinkles on it the life-giving plant and the other the life-giving water. Thus let Inanna arise."

Very interesting details: Enki sends torment on his daughter Erec-ki-gala like the suffering of a woman giving birth, and sends *gala-tura and the kur-jara* to announce to her that in exchange for ending her suffering, she should give up the corpse of Inanna. Gala-tura and the kur-jara, as non-human and ungodly beings (artificial, dead?), go unnoticed by Neti. Erec-ki-gala is trying to fight, seducing them with promises (and we learn that the attribute of the gods is the Word (magic spell?), people - Fate, and the underworld queen is the mistress of the Word and Fate). But the creatures follow Enki's instructions and in return conjure Erec-ki-gala (the following details are given in (Dedović, 2020) with the soul of heaven, earth, and sky (Here you can see the "global" Tree of Life, and, for example, the specifically Chinese San Tsai, chapter 8, etc.) Erec-ki-gala uses the last chance, trying to feed and drink them with food and water of the underworld (creating a "body of food" - annamayakosha and "biofield" - pranamayakosha, Chapter 13), and thus gain power over them, however, the beings reject these "gifts" as well. All they must do is fulfill the will of Enki:

[279-281] They were given the corpse hanging on the hook. One of them sprinkled on it the life-giving plant and the other the life-giving water. And thus, Inanna arose.

So, Inanna received the first two "shells". Now it is a kind of biorobot, a magical machine. The work of her resurrection is not yet completed, and the Anunna judges strictly monitor this, demanding the implementation of the laws of the underworld (Khazarzar, n / d):

Inanna emerges from the underworld.

The Anunna seize her.
"Which of those descending into the underworld
Came out unscathed
From the underworld?
If Inanna leaves the Land of No Return,
Let's leave the head behind her head!
Having finally received the "head" – nous, reason, will, ego – Inanna gives
them the "head" of her earthly husband (or lover) Dumuzi:
Bright Inanna gave her head to Dumuzi for her head!
Bright Ereshkigal!
Good song of praise for you!

Ultimately, Dumuzi's sister volunteers to replace herself, and all ends with her going to the realm of the dead for six months, and Dumuzi for six months.

There are analogies between the myth of Persephone, the voluntary sacrifice of Jesus Christ, and many similar plots in the myths of different peoples.

We conclude the comment by moving from Ars Magna to the concepts and terms of the theory of self-organization:

- A general post-mortem scenario is described - a convolution - however, the events occur in the reverse order (the author was embarrassed to expose the goddess, and only then remove the crown from her? Or is his deliberate encryption?);
- The return scenario is also described (even the need for an external impulse is indicated) - as a stratification in reverse order. A preserved body that does not interfere with activities in the underworld is a condition for return;
- And yet there are a few circumstances indicating that not a real death took place, but a deep trance, therefore, a resuscitation scenario is described.

The presence of different interpretations is a good reason to recall that the plot of a myth is not a sign, but a ternary connective, where an unambiguous interpretation is impossible in principle. Hence the fourth conclusion - an indication of a deep connection between the world of the living, the world of the dead, and the world of the gods - a real-world tree, stratifying and convoluting, along which souls wander in all three worlds.

The Path of Gilgamesh

The restless Ishtar (Inanna) tried to seduce Gilgamesh but ran into a rude refusal. After listing the sad fate of all her lovers, Gilgamesh stated (Dyakonov, 2006):

But I will not take you as my wife!
You are the brazier that goes out in the cold,
A black door that does not hold the wind and storm
The palace that crushed the head of the hero
The well that swallowed up its lid
The resin with which the porter is scalded,
The fur from which the porter is doused
A slab that did not hold back a stone wall,
The battering ram, who gave the inhabitants to the enemy land,
A sandal squeezing the master's foot!

In addition,

Enkidu heard these words of Ishtar,
He tore out the root of the bull, and threw it in her face:
"And with you - just to get it, as I would do as with him,
I would have wound his guts on you!"

For the goddess, this was already too much, and it is not surprising that the friends got into trouble, ending with the death of Enkidu[2].

Gilgamesh understand that the same fate awaited him and considered that since he was two-thirds a god, he was able to obtain immortality. And he decides:

To Uta-napishti, my father, I will go hastily,
To the one who, having survived, was accepted into the assembly of the gods and found life in it.
I'll ask him about life and death!

In search of Uta-napishti (Ziusudra), he follows the "path of the sun" (the sun god Shamash is called in the Epic a friend of Gilgamesh and Enkidu, (George, 2014):

To Mashu's twin mountains he came,
which daily guard the rising [sun,]
whose tops [support] the fabric of heaven,
whose base reaches down to the Netherworld.
There were scorpion-men guarding its gate,
whose terror was dread, whose glance was death,
whose radiance was fearful, overwhelming the mountains at
sunrise and sunset they guarded the sun.
Gilgamesh saw them, in fear and dread he covered his face,

> *then he collected his wits and drew nearer their presence.*
> *The scorpion-man called to his mate:*
> *'He who has come to us, flesh of the gods is his body.'*
> *The scorpion-man's mate answered him:*
> *'Two-thirds of him is god, and one third human.'* IX 38-50

So, the "path of the sun", like the "path of the earth", has its guardians, and these are scorpion people, semi-terrestrial, semi-subterranean border creatures, whose polymorphism, perhaps, indicates their Paleolithic origin (Chapter 13).

Gilgamesh managed to convince the scorpion man (undoubtedly, his friendship with Shamash and the fact that he was a descendant of the gods affected him):

> *The scorpion-man [opened his mouth to speak,]*
> *[saying a word] to King Gilgamesh, [flesh of the gods:]*
> *'Go, Gilgamesh! ....*
> *May the mountains of Mashu [allow you to pass!]*
> *'[May] the mountains and hills [watch over your going!]*
> *Let [them help you] in safety [to continue your journey!]*
> *[May] the gate of the mountains [open before you!]'* IX 130-135

Here we see a whole set of the border of the world's attributes - mountains, forests, gates, guards - well known from other myths. The scorpion-man opened the gates and talk about the further path.

> *Gilgamesh [heard these words,]*
> *what [the scorpion-man] told him [he took to heart,]*
> *he [took] the path of the Sun God ..... .*
> *At one double-hour ....,*
> *the darkness was dense, [and light was there none:]*
> *it did not [allow him to see behind him.]*
> *At two double-hours ....,*
> *the darkness was dense, [and light was there none:]*
> *it did not [allow him to see behind him.]*
> *At three double-hours ....,*
> *[the darkness was dense, and light was there none:]*
> *[it did not allow him to see behind him.]*
> *At four double-hours ....,*
> *[the darkness] was dense, [and light was there none:]*
> *it did not [allow him to see behind him.]*
> *At five double-hours ....,*

the darkness was dense, [and light was there none:]
it did not allow [him to see behind him.]
On [reaching] six double-hours,
the darkness was dense, [and light was there none:]
it did not allow [him to see behind him.]
On reaching seven double-hours ...,
the darkness was dense, and [light was there] none:
it did not allow him to see behind [him.]
At eight double-hours he was hurrying ...,
the darkness was dense, and light was [there none:]
it did not [allow him to] see behind him.
At nine double-hours ...... the north wind,
.... his face.
[The darkness was dense, and] light was [there none:]
[it did not allow him to] see behind him.
[On] reaching [ten double-hours,]
.... was very near.
[On reaching eleven double-hours a journey remained] of one double-hour,
[at twelve double-hours Gilgamesh came] out in advance of the Sun.
...... there was brilliance:
he went straight, as soon as he saw them, to ... the trees of the gods.
A carnelian tree was in fruit,
hung with bunches of grapes, lovely to look on.
A lapis lazuli tree bore foliage,
in full fruit and gorgeous to gaze on.
. . . cypress . ...
. . . cedar ....,
its leaf-stems were of pappardilu-stone and ...
Sea coral ...... sasu-stone,
instead of thorns and briars [there grew] stone vials.
He touched a carob, [it was] abashmu-stone,
agate and haematite ..... .
As Gilgamesh walked about ...,
she lifted [her head in order] to watch him. IX 145-195

So, Gilgamesh, "going forth by day" went through a tunnel in the bowels of the earth (remember Bosch and Moody, Chapter 10) and achieved what he wanted. There is a description of one of the types of transition from the world of the living to another world, known from many myths. But we need to clarify what kind of transition it was: was there a resuscitation, stabilization, or general post-mortem scenario? Or did this transition take place in a trance state?

The distances covered by Gilgamesh, as far as the surviving part of the text allows us to judge, are very similar in their properties. Therefore, this is really a designation of some "gaps" - the space and/or time of one, and not different transitions. Further, sensory deprivation takes place: *"the darkness is thick, no light is visible, neither forward nor backward can he see"*, and only at the end of the path do some tactile sensations appear, and then light. This is very reminiscent of the description of the stage of pratyahara before achieving samadhi (Chapter 10). There are also no descriptions of the internal dialogue, which can be compared with nirvitarka and nirvichara. Then the stage of mind purification from past impressions and complexes (vasanas and samskaras) should follow.

Gilgamesh meets the *"tavern-keeper of the gods"* Siduri:

> *Said the tavern-keeper to him, to Gilgamesh:*
> *'O Gilgamesh, there never has been a way across,*
> *nor since the olden days can anyone cross the ocean.*
> *Only Shamash the hero crosses the ocean:*
> *apart from the Sun God, who crosses the ocean?*
> *'The crossing is perilous, its way full of hazard,*
> *and midway lie the Waters of Death, blocking the passage forward.*
> *So besides, Gilgamesh, once you have crossed the ocean,*
> *when you reach the Waters of Death, what then will you do?*
> *'Gilgamesh, there is Ur-shanabi, the boatman of Uta-napishti,*
> *and the Stone Ones are with him, as he picks a pine clean in the*
> *midst of the forest.*
> *Go then, let him see your face!*
> *If [it may be] done, go across with him,*
> *if it may not be done, turn around and go back!'*
> *Gilgamesh heard these words,*
> *he took up his axe in his hand,*
> *and he drew forth the dirk [from] his [belt,].*
> *forward he crept and on [them] rushed down.*
> *Like an arrow he fell among them,*
> *in the midst of the forest his shout resounded.*
> *Ur-shanabi saw the bright . .....,*

> *he took up an axe, and he ... him.*
> *But he, Gilgamesh, struck his head ...,*
> *he seized his arm and ... pinned him down.*
> *They took fright, the Stone [Ones, who crewed] the boat,*
> *who were not [harmed by the Waters] of Death.*
> *......... the wide ocean,*
> *at the waters ... he stayed [not his hand]:*
> *he smashed [them in his fury, he threw them] in the river.* X 80-105

The classic description of catharsis followed by the enlightenment of the mind (chitta prasadanah). Let's pay attention to a monster at the bridge or at the crossing, which must be defeated before moving on - a typical mythological plot. It symbolizes the wild, sensual nature, which must be brought under control. Sometimes a person struggles with a monster until realizes that it is himself[3].

Describing what happened in terms of a generalized post-mortem scenario, we state that there is a convolution of the 5th-6th levels channels, and a partial convolution of the channel of the 4th level, which corresponds to the incomplete stage of transition 1. The transition must continue.

Gilgamesh meets Ur-shanabi and persuades him to cross over to Uta-napishti (you can't refuse Gilgamesh's charisma). Here's how the crossing happens:

> *Gilgamesh and Ur-shanabi crewed [the boat,]*
> *they launched the craft, and [crewed it] themselves.*
> *In three days they made a journey of a month and a half,*
> *and Ur-shanabi came to the Waters of [Death.]*
> *[Said] Ur-shanabi to him, [to Gilgamesh:]*
> *'Set to, 0 Gilgamesh! Take the first [punting-pole!]*
> *Let your hand not touch the Waters of Death, lest you*
> *wither [it!]*
> *'Take a second punting-pole, Gilgamesh, a third and a fourth!*
> *Take a fifth punting-pole, Gilgamesh, a sixth and a seventh!*
> *Take an eighth punting-pole, Gilgamesh, a ninth and a tenth!*
> *Take an eleventh punting-pole, Gilgamesh, and a twelfth!'*
> *At one hundred and twenty double-furlongs Gilgamesh had used all*
> *the punting-poles,*
> *so he, [Ur-shanabi,] undid his clothing,*
> *Gilgamesh stripped off [his] garment,*
> *with arms held aloft he made a yard-arm.* X 170-180

And again, there are parallels to many myths and rituals - for example, the burial of the dead, which took place among the ancient Scandinavians, especially the noble ones, in the ships set free by the waves and winds, which were also set on fire, or Noah's wanderings in the ark and so on.

Pay attention to the loss of the usual properties of space and time, the increase in their "possibility" - *"the path of six weeks in three days completed"*. These are "siddhis" already known to yogis - the ability to travel in space and time. There are also various smells and sounds.

Everything speaks of the completion of transition 1: the mind with its sensations of space and time is curtailed and reason and will are strengthened, becoming the main channels of interaction with the environment. Choice stage 2 follows, accompanied by riddles, and deceptions, revealing the true aspiration of a person - and his ability to comprehend the essence of things.

Gilgamesh, finally sailing to Uta-napishti, asks his questions:

> *Said Gilgamesh to him, to Uta-napishti the Distant:*
> *'I look at you, Uta-napishti:*
> *your form is no different, you are just like me,*
> *you are not any different, you are just like me.*
> *'I was fully intent on making you fight,*
> *but now in your presence, my hand stays.*
> *How was it you stood with the gods in assembly?*
> *How did you find the life eternal?'* XI 5

Uta-napishti tells the Sumerian version of the Deluge and begins testing Gilgamesh:

> *'But you know, who'll convene for you the gods' assembly,*
> *so you can find the life you search for?*
> *For six days and seven nights, come, do without slumber!'*

Gilgamesh slept for six days and seven nights, which for him passed in a moment. Here is the new task:

> *Take him, Ur-shanabi, lead him to the washtub,*
> *And have him wash his matted locks as clean as can be!*
> *'Let him cast off his pelts, and the sea bear them off,*
> *let his body be soaked till fair!*
> *Let a new kerchief be made for his head,*
> *let him wear royal robes, the dress fitting his dignity!* XI 255

Gilgamesh, taking the symbol for a sign, and the image for an instruction, follows it word for word - and receives unbearable clothes instead of an immortal body. Notice how the Waters of Death become the Waters of Life.

Uta-napishti gives the last opportunity:

Let me disclose, 0 Gilgamesh, a matter most secret,
to you [I will] tell a mystery of [gods.]
'There is a plant that [looks] like a box thorn,
it has prickles like a dogrose, and will [prick one who plucks it.]
But if you can possess this plant,
[you'll be again as you were in your youth.],
Just as soon as Gilgamesh heard what he said,
he opened a [channel] ..... .
Heavy stones he tied [to his feet,]
and they pulled him down ... to the Ocean Below.
He took the plant and pulled [it up, and lifted it,]
the heavy stones he cut loose [from his feet,]
and the sea cast him up on its shore. XI 270-290

Success - but Gilgamesh did not want to live forever alone: he wanted to deliver a plant to all the inhabitants of Uruk - and on the way the serpent stole the flower.

The one who makes the wrong choice, who could not give up his past, is not worthy of eternal life - Gilgamesh learns this lesson, accepts his mortality, returns to his hometown, and "makes a name for himself", building walls and telling poets about his campaigns. Well, we still remember him. So:

- Convolution completed at selection stage 2 (testing).
- Stabilization scenario, stage of certainty 4: Gilgamesh did not succeed, but Uta- napishti succeeded.
- The reason for the failure was Gilgamesh's misunderstanding of Uta-napishti's advice on how to create a new body.
- For Gilgamesh, there was a resuscitation scenario (return), the key to which was the "earthly" body, which, having protection with the help of amulets, actively acted in the underworld.

ANCIENT EGYPT: THE WAY OF GOING FORTH BY DAY

It is customary to give an enthusiastic assessment of the Egyptian Book of the Dead (Shaposhnikov, 2003). While remaining objective, the following should be noted that the language and symbolism of the Egyptian Book of the Dead are very complex, the presence of dialogues and different opinions in the text does not make it easier to understand, and the construction of the ternary connective encounters some dark places. Let us give the Book of the Dead a visual form, a sampler for understanding (Figures 7-9).

FIRST 70 DAYS AFTER DEATH

Figure 7. First 70 days after death. Events are described in orange frames according to the Book of the Dead. Used: translation of the Book of the Dead and commentaries

First 70 days after death	
Human Components	Actions of Priests and Paraschites
Separation of Khet and (Ka, Ba, Ib 1-3, Ah) flying off through the mouth of the deceased	Rite of search for the Eye of Ujat
	Mummification
Wandering (Ka, Ba, Ib 1-3, Ah) in space visiting the Moon, Planets and Sun	
Periodic return (Ka, Ba, Ib 1-3, Ah) to Khet in order to monitor the correctness of the performance of the rites. If the rites are performed carelessly, the Ka of the deceased will be offended and will turn into an evil twin spirit (ghost) that will forever haunt his family, sending disasters on the descendants.	

After 70 days, the funeral procession, led by the priests in the clothes and masks of the gods of the Duat, swam across the Nile, landed on the west bank and approached the tomb. At the entrance the coffin was placed on the ground and the "gods of the Duat" performed the rite of "opening the mouth", returning to the deceased his Ba and creating his Sah. The deceased regained the ability to eat, drink and speak. The goddesses associated with childbearing (Isis et al.), the scarab amulets and the Eye of Ujat, took part in this "second birth". Then the priests placed the coffin in the sarcophagus. Near the walls of the burial chamber, canopies were placed. Amulets and figurines of the air god Shu were installed so that the deceased would not suffocate in the Underworld. Four amulets were placed in the walls, driving away evil spirits, four lights were lit. The door of the crypt was sealed with the seal of the necropolis, laid with blocks and covered with rubble.

(Granin, 2019; Goelet, et al., 2015; Rak, 1993; Shaposhnikov, 2003)

Assessing the actions of the priests within the framework of the general posthumous scenario, we note that, to keep the body, they try to save the structural symmetry of the wave and soliton components (Chapter 11), so that a human who appears before the gods, had maximally preserved earthly abilities. In addition to the wave component that preserves the levels and channels of the subjective space, the priests create an artificial soliton, which becomes a "body" adapted to "life" in the world of the dead, thereby starting the restoration of the soliton-wave system - this process is completed after Osiris endows this system Ah (Hu).

FROM THE GATE OF THE HOUSE OF OSIRIS TO THE HALL OF TWO TRUTHS

Figure 8. Path to the Hall of Two Truths

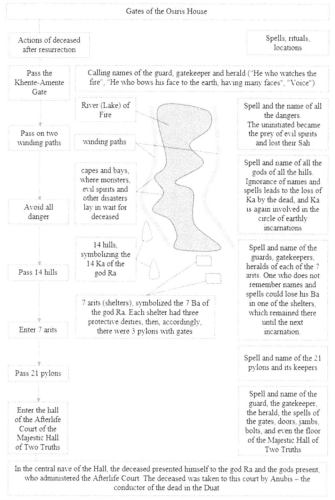

(Granin, 2019; Goelet, et al., 2015; Rak, 1993; Shaposhnikov, 2003)

The Egyptians, trying to help the dead in everything, depicted this path on the floor of the coffin (Figure 9, right top corner).

Figure 9. A detail from the floor of a coffin of Gua, physician of Djehutyhotep, a nomarch of Deir el-Bersha, Egypt, during the Middle Kingdom

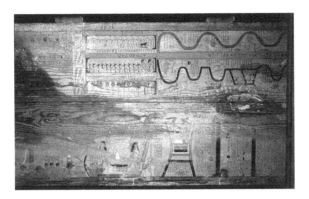

(Tann, 2019)

AFTERLIFE COURT. THE ULTIMATE FATE OF THE DEAD

Let us citate (Shaposhnikov, 2003) and comment on him within the frame of post-mortem scenarios (Chapter 11):

Among the gods, there were certainly the gods of the Great Host (Ra, Shu, Tefnut, Geb, Nut, Horus, Isis, Nephthys, Hathor, Set, the god of divine will Hu, and the god of reason Sia) and the gods of the Lesser Host (42 gods, according to the number of nome regions Egypt). The deceased make a speech of justification before each of the hosts. The assembly of the gods included the Great Ennead of Heliopolis, which included Ra-Atum and eight gods, originating from Atum: Shu and Tefnut, Geb and Nut, Nephthy,s and Set, Isis, and Osiris; the Memphis Triad: Ptah, Sekhmet, and Nefertum; The Great Ogdoada, that is, the "eight" of gods who personified the elements: male deities with frog heads and their female pairs with snake heads - Khukh and Hauket, Nun and Naunet, Amon and Amaunet, Kuk, and Kauket; Small Ennead (42 gods). Having uttered a greeting to Ra and the gods, the deceased in his "confession of denial" renounce all sins. He lists 42 sins and swores to the gods that he had not committed them and was not guilty.

Meanwhile, Thoth and Anubis were weighing the dead Eb on the Scales of Two Truths. Ib was placed on one bowl, and the feather of the goddess of justice Maat was placed on the other. If the deceased lied, denying his sin, the pointer of the scales deviated. If the scales remained in balance, the deceased was recognized as "right-handed."

After the confession of the denial of all sins, the deceased had to turn to the Small Host of the gods, call each of the 42 gods by name and deliver the "second justification".

When the interview ended, the guardian spirit Shai, the goddess Meskhent, the goddess of good fortune Renenutet, and the soul-manifestation of the deceased Ba came forward. They testified to the character of the deceased, his good and evil thoughts, words, and deeds. The goddesses Neith, Nephthys, Isis, and Serket the scorpion spoke in defense of the deceased. Taking into account the results of the weighing of the soul-Ib by Thoth and Anubis, the evidence of the soul-Ba of the deceased, the arguments of the accusers Meskhent, Shai, Renenutet, and the defenders of Isis, Nephthys, and Neith, the Great Ennead pass a verdict.

If the verdict was guilty, then the heart of the deceased was given to be eaten by the terrible goddess Amma (Eater), a monster with the body of a hippopotamus, lion's paws, a mane, and a mouth of a crocodile. If the verdict turned out to be an acquittal, all the shells of the deceased were sent to the Holy of Holies of the Temple of Two Truths, to the throne of Osiris.

The deceased kissed the threshold, blessed the parts of the door and the nave of the Sanctuary, and appeared before Osiris, who sat surrounded by Isis, Nephthys, Maat, the clerk Thoth, and the four sons of Horus in a lotus flower. He announced the arrival of the deceased, cleansed of all sins, and numbered among the saints. After a gracious conversation with the visitor, the gods sent him to the Abode of Eternal Bliss (Reed Field, or Field of Satisfaction), accompanied by the guardian spirit Shai. The way to the abode of blissful spirits (Ah) was blocked by the last Gate, which should be called by name and cast spells by the guardian god.

Consider the symbolism of the path traveled by the deceased (Shaposhnikov, 2003):

Overcoming the First Gates of the Osiris House means final death, and the impossibility of passing them implies a way out of clinical death, or stupor, coma, or lethargic sleep. This is a barrier only for the physical body - Khet; souls overcome it.

Let us clarify, that resuscitation is not possible immediately after the start of mummification - procedures such as brain pulling or body gutting are not compatible with life. Here, the initial stage 0 of the generalized post-mortem scenario is not described.

All the preliminary wanderings of the shells in space, their sojourn on the Sun and the Moon, is a description of the stage of transition 1 (Chapter 11), symbolized by the crossing of the funeral procession across the Nile. The stage of choice 2 begins, and it marks the entrance to the First Gate of the House of Osiris.

The first series of trials (two Paths near the Lake of Fire) had as its goal either to return the soul-body (Sah) of the deceased to the circle of earthly incarnations or to save him from them. The dangers and monsters that seize Sah symbolize the return of Sah to the earthly world, that is, the acquisition of a new body. Successful passage of the paths saves Sah from a new incarnation in bodily form.

The decomposition of the Hut body is described here, after which either re-incarnation scenario 2 is possible (here Sah, holding all the shells, is used as a matrix at the birth of a new physical body Khet), or a continuation of the general posthumous scenario.

The second series of trials (14 hills and their deities) aims to determine the future fate of Ka. Failure plunged Ka into a new incarnation on Earth. Successful passage saved Ka from returning to the earthly vale.

Alternative: the wave part of the system either forms a "symmetric" soliton (in the act of conception, Chapter 6), forming a soliton-wave system or continues convolution.

The third series of trials (7 arts and 21 pylons) was intended to decide the fate of the soul-manifestation of Ba. If the souls of the deceased failed to enter one of the shelters, then Ba remained in the previous arita. She was doomed to move into something material on one of the seven sacred luminaries (Sun-Earth, Moon-Earth, Saturn, Venus, Jupiter, Mercury, and Mars). Successful passage of 7 arits freed Ba from moving into bodies on the planets.

In terms of the theory of self-organization, a ternary connective is being built, where Ba plays the role of a projection that provides the choice of the place and time of conception if the test is not passed. The planets can be interpreted in an astrological sense, but perhaps extraterrestrial birthplaces are also meant. If the test is passed, Ba, also as a projection, provides the correct direction for the transition stage 3 to the certainty stage 4 (Chapter 11) with the appropriate convolution. Perhaps this is personified by the metapomp - Anubis.

The fourth trial (the Judgment after death, the weighing of Eb) determined the fate of the three energy shells.

Transition stage 3 ends. Certainty stage 4 begins.

The allegorical devouring of Ib by the monster Amma means the return of Ib to the circle of transmigration of souls, most likely, a connection with Ba, Ka, and Sah, who lagged at the previous stages of testing.

Without any allegory: this is a failure of the stabilization scenario - instead of establishing exchanges with the environment as an open system, the wave becomes "food" for other waves or breaks into solitons (variants of "unsuccessful" superposition, interference, and diffraction, considered in Chapter 11). This is, as defined in Chapter 11, the ultimate death scenario.

A successful entry into the Holy of Holies of the Temple of Two Truths marks the deification of the deceased, the likening of his spirit (Ah) to the blessed spirits of the gods (the same Ah). The spirit can remain among the related spirits of the Host of the Gods and become an accomplice in their existence.

This is a successful implementation of the stabilization scenario. We see the weakening of the will and reason, the preservation of the ego, and the strengthening of intuition (a sense of the kinship of spirits). Probably, it is here that the spiritual body of Sah is finally formed, receiving the resources of existence no longer in the form of a magical "charging", but because of an established interaction with the environment - as a gift from Osiris according (Granin, 2019):

If the spirit Ah has overcome the Last Gate on the way to the Field of Satisfaction, then it will gain eternal bliss, will no longer know the torments of moving from body to body, and get rid of the incarnation suffering.

Last convolution and merge stage 5.

It seems. that the Egyptian initiates also indicated the habitat of blissful souls - the Sun[4].

However, complete merging with the universal spirit Ah (Hu), with the loss of even the ego, could satisfy only highly spiritual people. For the people:

Egyptian folk ideas about what kind of life awaited the deceased on the Field of Reeds have nothing to do with mystical experience. It was impossible to expect anything else because only parables are possible about the inexpressible. One of them is the description of the Reed Field. There, the deceased is allegedly waiting for the same life that he led on Earth, only freed from suffering and trouble, happier and better. His spirit will lack nothing. Seven Hathor, the god of grain Nepri, the scorpion-Serket, and other gods will make his arable land and pastures exceptionally fertile, fat herds, numerous and fat birds. All work in the fields will be done by ush-abti workers. So, in satiety, contentment, love joys, singing, and dancing, millions of years of blissful existence will pass. This mundane ideal appealed to the common people and slaves, and the initiates did not seek to refute it.

The same apotheosis of ordinary consciousness, as in Table 12 ("Enkidu in the underworld"), supplementing the Epic of Gilgamesh. Where the initiates saw a symbol and created a ternary connective, ordinary people saw a sign and a literal description. It can be confidently asserted that the attempts of the initiates to explain themselves would lead to their immediately being torn to pieces by the crowd, deceived in their ideals.

To summarize: the levels and channels, their states and sensations, the stages, and transitions of the convolution of the SWM are invariant to the shells, sensations and speeches, trials, and events of the Book of the Dead. Both the SWM and the text of the Book of the Dead are projections in one ternary connective. So, the general post-mortem scenario and special cases are confirmed by the knowledge of ancient Egyptians.

(Shaposhnikov, 2003) considers the mystical experience of initiated priests to be the source of such detailed knowledge. In addition, let us recall that pre-dynastic Egypt, as the historical Egyptians already believed, was ruled by gods, demigods, and then by the followers of Horus, that is, the source could be divine revelation. The memory of the "common human heritage" of the Stone Age, from which the ancient Egyptians did not go too far, also was strong.

THE DEATH OF ADAM AND THE LIFE OF JESUS

Let's look at the Abrahamic religions only from the standpoint of convolution/stratification scenarios, which, of course, is theologically unsatisfactory, but it meets the objectives of this research.

So, Adam and Eve, not realizing that all the good things they had in Paradise, they owe not to its beauties or the usefulness of the fruits, or even knowledge, but exclusively to the presence of God, relied on their practical mind, sinned and, feeling changes in their state, tried to "flee the scene of the crime." And, since even the leading questions of God did not cause a change in their way of thinking (repentance, metanoia), God did what their innermost desire was - he delivered Adam and Eve from his presence. He even did more: he gave them leather garments (bodies) adapted for life in a world with different conditions.

This is the separation of the soul from God, according to St. Gregory Palamas (St. Gregory Palamas, 2011)., and there is death: *And God warned them not to eat from the Tree of the Knowledge of Good and Evil, saying, turning to Adam: for on the day, you eat from it, you will die by death.*

Thus, it is exactly as death that the expulsion of the forefathers from paradise should be considered, the gates of which, guarded by an Angel with a fiery sword, closed for them forever - just as the world of the living forever becomes inaccessible for the dead.

At the same time, the forefathers did not completely lose the ability to communicate with God, nor did they lose their ego, will, and reason - but these channels ceased to be dominant, yielding to the practical mind seeking benefits. Let us now emphasize the following facts:

- the excessive development of the practical mind brought Adam and Eve into the world with time and space (channel of the 4th level, the objects of perception of which are space and time, Chapter 8);
- this is interpreted as a stratification from levels 1-3 to level 4;
- as it should be during a stratification, the potential of the previous levels decreases (but does not reset to zero) (weakening of intuition, ego, will, and reason), which is partially compensated by external exchanges (and you will eat field grass). The corresponding formulas and models are given in chapters 3 and 5;

- movement from world to world is not only a physical movement, but also a change in consciousness, and this is carried out by death, which directly follows from the abundant ethnographic material: for example, the movement of boys from the world of childhood to the world of adult men is ritually played out as death, and the death throes trying to be as realistic as possible. In Christianity, baptism is carried out by ritual immersion in the waters of death, which become the living waters of new life (such ambivalence of waters was present even in the Epic of Gilgamesh). So, death here is not a figure of speech, as modern man understands it, but a literal death, which is emphasized by the ritual;
- leather garments belong to the world of space and time, they, that is, the body, die, and the spirit and soul (in the understanding of chapter 13), which do not belong to such a world, remain to exist (but may disappear as a result of convolution);

The very act of being, transferred to the new world, is a synergy of the will of God and the will of Adam and Eve.

Adam and Eve die, their long-lived descendants die, and then the following happens (Holy Bible. (ASV):

> 5 And Jehovah saw that the wickedness of man was great in the earth, and that every imagination of the thoughts of his heart was only evil continually.
> 6 And it repented Jehovah that he had made man on the earth, and it grieved him at his heart.
> 7 And Jehovah said, I will destroy man whom I have created from the face of the ground; both man, and beast, and creeping things, and birds of the heavens; for it repenteth me that I have made them.
> 11 And the earth was corrupt before God, and the earth was filled with violence.
> 12 And God saw the earth, and, behold, it was corrupt; for all flesh had corrupted their way upon the earth.
> 13 And God said unto Noah, The end of all flesh is come before me; for the earth is filled with violence through them; and, behold, I will destroy them with the earth.

(Gen. 6)

From the description of the deeds of the descendants of Adam, it is quite clear that wickedness is understood as the development of strong passions that the practical mind (nus) could no longer control, concentrating at all times on evil thoughts. These passions became incompatible with the world of Adam and Eve, or rather, with that "remoteness" from God that was characteristic of this world.

And then the waters of the Flood (the element of water is associated with feelings and emotions) cover and destroy the old world.

Only Noah and his family were able to live in the world of passions - and were transferred in the act of death (and again we can talk about the synergy of the will of God and man) into the modern world of short life, with a corresponding weakening of intuition, ego, will, reason and nous.

This is interpreted as a stratification up to levels 5 and 6.

There are several episodes in the Bible when a change in the state of consciousness is carried out through ritual death, for example, the Abraham exodus from Ur of the Chaldees (he received the necessary qualities to become the ancestor of God's chosen people, his ego decreased, intuition increased), the exodus of the Jews from Egypt and forty years wanderings in the desert (children of fearful refugees became aggressive and cruel conquerors - ego, will, mind increased). Here, death corresponds to transitions between stages of convolution (with excessive compensation for potential - this is how God's assistance is expressed), and a new stage of convolution starts a new life.

A few similar episodes are contained in the path of Jesus - baptism, after which He changed the life of a carpenter's son to the life of a preacher, a forty-day fast in the desert, after which He rejected all the temptations of the devil, and gain the life of a saint, and finally, voluntary self-sacrifice - death on the cross, which changed the fate of mankind. All of them, except for the ritual, also had a representative value, becoming examples for followers.

The last of these episodes is the most important. Why Jesus is called the firstborn of the dead in the Announcement for Easter by St. John Chrysostom? The prophets and Jesus himself resurrected the dead even before that. Let's analyze the meaning of this episode in more detail.

So, according to Church tradition, God promised to save the descendants of Adam from death by returning them to paradise. Since the expulsion from paradise was a synergy of the wills of God and man, their salvation should have been caused by synergy.

At the right moment, the family of Noah was chosen from the antediluvian mankind, then the Jewish people became God's chosen people, then the King David family, and, finally, the Virgin Mary turned out to be worthy to become the mother of the God-man. The strictest millennia-old selection, which created the unity of God's will and man's will, voluntarily following the will of God.

At the same time, the human will by no means lose its independence in the dramatic prayer episode on the Olives Mount, when the whole nature of a person was horrified by the impending painful death, but voluntarily agreed to sacrifice itself.

So, Jesus, as a person, consciously performs an act of humility - in the optics of the theory of self-organization, this means convolution of higher channels - ego, will, and reason.

This is followed by representative actions that are, as it were, "skew-symmetric" to the events of the Egyptian Book of the Dead. Thus, the humiliation and beatings in the courtyard of the high priest by the people of the God-man, who was the source of life, are comparable to the trials around the Fiery Lake Sah and Ka of the deceased; the judgment by the Sanhedrin and Pilate, who may have read the Sibylline books with prophecies about the Savior - with the Judgment Beyond the grave by the gods, and if in the second case, Ib of the deceased was tested, then in the first, in fact, the hearts of the people, priests, and Pilate.

And, finally, the culmination: the curse of the God-man, (cursed is everyone, hang on a tree), his death on the cross. The words Eli, Eli, lama sabachthani? fix the moment of the separation of the Spirit of God from the spirit of man, without which death is impossible, followed by the separation of the soul from the body, that is, death as the destruction of an integral soliton-wave system. This is comparable to the devouring of Ib of the deceased by the monster Amma - that is the destruction of the integral system of shells. But how different ("skew-symmetric") consequences! In the second case, this is the failure of the stabilization scenario and final death, but in the first case, the destruction of the gates of Hell, the liberation of the righteous, resurrection, paradise, sitting at the right hand of the Father and changing the fate of Adam's descendants - now paradise can again be gained by them. Here it is appropriate to quote the already mentioned Paschal Sermon of St. John Chrysostom in full (St. John Chrysostom. Paschal Sermon, n / d):

If any man is devout and loves God, let him enjoy this fair and radiant triumphal feast. If any man be a wise servant, let him rejoicing enter into the joy of his Lord. If any have labored long in fasting, let him now receive his recompense. If any have wrought from the first hour, let him today receive his just reward. If any have come at the third hour, let him with thankfulness keep the feast. If any have arrived at the sixth hour, let him have no misgivings; because he shall in nowise be deprived thereof. If any have delayed until the ninth hour, let him draw near, fearing nothing. If any have tarried even until the eleventh hour, let him, also, be not alarmed at his tardiness; for the Lord, who is jealous of his honor, will accept the last even as the first; he gives rest unto him who comes at the eleventh hour, even as unto him who has wrought from the first hour.

And he shows mercy upon the last, and cares for the first; and to the one he gives, and upon the other he bestows gifts. And he both accepts the deeds, and welcomes the intention, and honors the acts, and praises the offering. Wherefore, enter you all into the joy of your Lord; and receive your reward, both the first and likewise the second. You rich and poor together, hold high festival. You sober and you heedless, honor the day.

Rejoice today, both you who have fasted and you who have disregarded the fast. The table is fully-laden; feast ye all sumptuously. The calf is fatted; let no one go hungry away.

Enjoy ye all the feast of faith: Receive ye all the riches of loving-kindness. let no one bewail his poverty, for the universal kingdom has been revealed. Let no one weep for his iniquities, for pardon has shown forth from the grave. Let no one fear death, for the Savior's death has set us free. He that was held prisoner of it has annihilated it. By descending into Hell, He made Hell captive. He embittered it when it tasted of His flesh. And Isaiah, foretelling this, did cry: Hell, said he, was embittered when it encountered Thee in the lower regions. It was embittered, for it was abolished. It was embittered, for it was mocked. It was embittered, for it was slain. It was embittered, for it was overthrown. It was embittered, for it was fettered in chains. It took a body and met God face to face. It took earth and encountered Heaven. It took that which was seen, and fell upon the unseen.

O Death, where is your sting? O Hell, where is your victory?

Christ is risen, and you are overthrown. Christ is risen, and the demons are fallen. Christ is risen, and the angels rejoice. Christ is risen, and life reigns. Christ is risen, and not one dead remains in the grave. For Christ, being risen from the dead is become the first fruits of those who have fallen asleep. To Him be glory and dominion unto ages of ages. Amen.

Now let's describe in more detail the Christian version of what happens with the soul after death and before the Last Judgment, and compare it with the general post-mortem scenario. Little is said about this in the Holy Scriptures, therefore Church Tradition, the texts of the funeral services, and texts of contemporary authors will be used (Vasiliadis, 2012; Prayers for the Dead, 2016; The Ordeals of St. Theodora, n / d)[5].

We note right away that in Christianity bipolarity (the channel of reason) is traced: angels and demons, hell and heaven, however, God doesn't violate the free will of His creatures which choose good (synergy with the will of God) or evil (deviation from synergy).

Consider the near-death and post-mortem events point by point.

Seeing The Spirits During the Dying

They see things that the living do not see - saints, angels, deceased friends, relatives, and demons (Chapter 10). We quote (The Ordeal of St. Theodora, n / d) - the story is told on behalf of St. Theodora:

So, when the hour came for the separation of my soul from the body, I saw around my bed a lot of Ethiopians, black as soot or pitch, with eyes burning like coals. They raised noise and shouted: some roared like cattle and beasts, others barked like dogs, others howled like wolves, and others grunted like pigs. All of them, looking at me, raged, threatened, gnashed their teeth, as if they wanted to eat me; they prepared charters in which all my bad deeds were recorded. Then my poor soul trembled; it was as if the torment of death did not exist for me: the terrible vision of the terrible Ethiopians was for me another, more terrible death. I turned away my eyes so as not to see their terrible faces, but they were everywhere and their voices were carried from everywhere. When I was completely exhausted, I saw two Angels of God approaching me in the form of beautiful youths; their faces were bright, their eyes looked with love, the hair on their heads was white as snow and shone like gold; the clothes were like the light of lightning, and on the chest, they were cross-girded with golden belts. Approaching my bed, they stood beside me on the right side, talking quietly to each other.

When I saw them, I rejoiced; the black Ethiopians trembled and moved away; one of the bright young men addressed them with the following words: "O shameless, accursed, gloomy, and evil enemies of the human race! rejoice very much, here you will not find anything, for God is merciful to her and you have no part and share in this soul. After listening to this, the Ethiopians rushed about, raising a strong cry and saying: "How can we not have a part in this soul? and this?" And having said this, they stood and waited for my death. Finally, death itself came.

A typical picture of near-death experiences. Theodora perceives them as pre-death experiences.

The First Two Days After Death

The soul can visit those places on earth that are dear to her, but on the third day, she moves to the afterlife. Tradition reports that the angel who accompanied St. Macarius of Alexandria explained the church commemoration of the dead on the third day (St. Macarius of Alexandria, 1831): *for two days the soul, along with the Angels who are with it, is allowed to walk the earth wherever it wants. Therefore, the soul that loves the body sometimes wanders near the house in which it was separated from the body, sometimes near the tomb in which the body is laid; and thus spends two days, like a bird, looking for its nests. And a virtuous soul walks into those places where it used to do the right thing.*

In (The Ordeal of St. Theodora, n / d) the state after death and the need for commemoration and support are described as follows:

But the Holy Angels began to look for my good deeds and, by the grace of God, they found and collected everything that with the help of the Lord was well done by me: whether I ever gave alms, or fed the hungry, or gave the thirsty to drink, or clothed the naked, or led the stranger into her house and calmed him, ... and all my other smallest deeds were collected by holy angels, preparing to lay against my sins. The Ethiopians, seeing this, gnashed their teeth, because they wanted to kidnap me from the Angels and take me to the bottom of hell.

At this time, our reverend father Basil unexpectedly appeared there and said to the Holy Angels: "My Lord, this soul served me a lot, calming my old age, and I prayed to God, and He gave it to me." Having said this, he took out a golden bag from his bosom, all full, as I thought, of pure gold, and gave it to the holy angels, saying: "When you go through air ordeals and evil spirits begin torturing this soul, redeem her with this from her debts. I am rich by the grace of God because I have collected many treasures for myself by my labors, and I give this bag to the soul that served me. Having said this, he disappeared. The crafty demons, seeing this, were perplexed and, raising lamentable cries, also disappeared. Then the saint of God, Basil, came again and brought many vessels with pure oil, dear ointment, and, opening each vessel one by one, poured everything on me, and a fragrance spilled from me. Then I realized that I had changed and become especially bright. The saint again turned to the angels with the following words: "My Lord! When you have done everything necessary for her, then, having brought her to the monastery prepared for me from the Lord, leave her there."

In the Orthodox rite of burial of the departed Ven. John of Damascus describes the state of the soul that has parted from the body, but is still on earth, as follows: *Alas! What an agony the soul endures when from the body it is parting; how many are her tears for weeping, but none will show compassion: unto the angels, she turns with downcast eyes; useless are her supplications; and unto men, she extends her imploring hands, but finds none to bring her rescue. Thus, my beloved brethren, let us all ponder well how brief the span of our life; and peaceful rest for him (her) that now is gone, let us ask of Christ, and also His abundant mercy for our souls.* (The Funeral Hymns of St. John of Damascus. 2008).

Here we note the following:

- Possibility of a resuscitation scenario under certain conditions: prayer for the deceased and God's will – almost all the cases of the resurrection of the dead mentioned in the Bible took place immediately after death.
- The role of church rites and individual prayers for the departed (St. Basil's bag of gold) in alleviating their posthumous fate (denied by Protestants).

- 2 days - corresponds to the initial stage 0 of the generalized post-mortem scenario, if there was no resurrection.

On The Third Day

In the same revelation of the Angel, St. Macarius of Alexandria says this: *On the third day, He Who rose from the dead, commands, in imitation of His resurrection, to ascend to heaven for every Christian soul to worship the God of all.*

On the third day, it is customary to bury the dead. The Church makes an offering and a prayer for his soul. Thus, day 3 corresponds to transition stage 1.

From The Third Day to The Ninth

Tradition says that *after worshiping God, He is commanded to show the soul the various and pleasant abodes of the saints and the beauty of paradise. All this is considered by the soul for six days, wondering and glorifying the Creator of all this - God. Contemplating all this, she changes and forgets the sorrow she had while in the body. But if she is guilty of sins, then at the sight of the pleasures of the saints, she begins to grieve and reproach herself, saying: "Alas!" to me! How I fussed in that world! Carried away by the satisfaction of lusts, I spent most of my life in carelessness and did not serve God as I should, so that I could also be rewarded with this goodness <...> After considering all the joys of the righteous for six days, she again ascends by angels to worship God. So, the Church does well, making services and offerings for the deceased on the ninth day.*

After consideration for six days of all the joys of the righteous, she again ascends by angels to worship God. And so, the Church does well, making services and offerings for the deceased on the ninth day (St. Macarius of Alexandria, 1831).

Ordeals From Ninth to The Fortieth Day

After the second worship, the Lord of all again orders to take the soul to hell and show it the places of torment located there, the different sections of hell, and the various wicked torments. Through these various places of torment, the soul rushes for thirty days, trembling, so as not to be condemned to imprisonment in them (St. Macarius of Alexandria, 1831).

St. Theodora, on the one hand, does not single out these periods, and on the other hand, describes in detail the so-called ordeal tortures, which are not mentioned in the previous fragment (The Ordeal of St. Theodora, n / d):

The holy angels took me from the earth, and went up to heaven, ascending, as it were, through the air. And so, on the way, we suddenly met the first ordeal, which is called the ordeal of idle talk and foul language. Here we stopped. We were brought out a multitude of scrolls, in which were written down all the words that I had only spoken from my youth, all that I had said thoughtlessly and, moreover, shameful. All the blasphemous deeds of my youth were written down, as well as the idle laughter to which youth is so prone. I immediately saw the bad words that I had ever spoken, shameless worldly songs, and the spirits denounced me, pointing out both the place and the time and the persons with whom I engaged in idle conversations and angered God with my words, and did not consider him a sin at all. and therefore, did not confess this to the spiritual father. Looking at these scrolls, I was silent as if deprived of the gift of speech, because I had nothing to answer them: everything that was written down by them was true. And I was surprised how they didn't forget anything because so many years have passed and I myself have long forgotten about it. They tested me in detail and in the most skillful way, and little by little I remembered everything. But the holy angels who led me put an end to my trial at the first ordeal: they covered my sins, pointing out to the evil ones some of my former good deeds, and what was missing from them to cover my sins, added from the virtues of my father, the Monk Basil, and redeemed me from the first ordeal, and we went further.

Then follow the ordeals of Lies, Condemnation and Slander, Overeating and Drunkenness, Laziness, Theft, Love of Money and Avarice, Covetousness, Untruth, Envy, Pride, Wrath, Resentment, Robbery, Sorcery, charm, poisoning, invocation of demons (the spirits of this ordeal in their form similar to four-legged reptiles, scorpions, snakes, and toads; is it an accidental resemblance to the scorpion-man met by Gilgamesh?), Fornication, Adultery, Sodomy, Idolatry and all sorts of heresies, Unmercy and Hardness of Heart.

Along the way, the Angels explain to Theodora that *every Christian from the very Holy Baptism receives from God a Guardian Angel, who invisibly guards a person and throughout his life, even until the hour of death, instructs him in all good and all these good deeds that a person does during his life. earthly life writes it down so that he can receive mercy them from the Lord and eternal recompense in the Kingdom of Heaven. So, the prince of darkness, who wants to destroy the human race, assigns to each person one of the evil spirits, who always walks after a person and watches all of his youth, and evil deeds, encouraging them with his machinations, and collects everything that a person has done wrong. Then he refers to all these sins as ordeals, writing each in the appropriate place. From here all the sins of all the people that only live in the world are known to the airy princes. When the soul is separated from the body and strives to ascend to heaven to its Creator, then the evil spirits hinder it, showing the lists of its sins; and if the soul has well more than sins, they cannot keep her; when there are more sins on her than good deeds, then they hold her for a while, imprison her in the ignorance of God and torment her, as far as the power of God allows them, until the soul, through the prayers of the Church and relatives, receives freedom. If, however, it turns out that a soul is so sinful and unworthy before God that all hope for its salvation is lost and it is threatened with eternal death, then it is brought down into the abyss, where it remains until the second coming of the Lord, when eternal torment in fiery hell begins for it. Know also that only the souls of those who are enlightened by holy baptism are tested in this way. But those who do not believe in Christ, idolaters, and in general all those who do not know the true God do not ascend this way, because during earthly life they live only in body, but in soul they are already buried in hell. And when they die, demons without any trial take their souls and bring them down to hell and the abyss.*

For the souls of believers ascending to heaven, there is no other way - everyone goes here. But if someone sincerely repents of all sins, then the sins, by the mercy of God, are invisibly blotted out, and when such a soul passes here, the airy tormentors open their books and find nothing written behind it; and the soul ascends in joy to the throne of grace.

According to Tradition, when the Archangel Gabriel informed the God Mother of the death approach, She prayed to His Son to deliver Her soul from these demons, and the Lord Jesus Christ appeared from heaven to accept His Mother's soul and take Her to heaven.

It is easy to see that ordeals are the results and projections of non-compliance with the Commandments of Moses, defeats in the fight against seven (or eight, if you read vanity and pride separately) main passions, violation of the Commandments of Love of Jesus Christ. In general, the Orthodox List contains 377 mortal and non-mortal sins.

Further, St. Theodora and the Angels accompanying her approach the heavenly gates:

With this, a series of aerial ordeals ended, and we joyfully approached the gates of heaven. These gates were bright as a crystal, and all around there was a radiance that cannot be described; sunlike young men shone in them, who, seeing me led by angels to the heavenly gates, were filled with joy because I, covered by the mercy of God, went through all the aerial ordeals. They kindly greeted us and led us inside.

What I saw and what I heard there - it's impossible to describe! I was brought to the Throne of God's impregnable glory, which was surrounded by Cherubim, Seraphim, and multitudes of heavenly armies, praising God with unspeakable songs; I fell on my face and bowed to the invisible and inaccessible to the mind of the human Deity. Then the heavenly powers sang a sweet song, praising the mercy of God, which the sins of people cannot exhaust, and a voice was heard commanding the angels who led me to take me to see the abodes of the saints, as well as all the torments of sinners, and then calm me down in the abodes prepared for the blessed Vasily. By this command, I was taken everywhere, and I saw villages and cloisters filled with glory and grace, prepared for those who love God. Those who led me showed me separately the cloisters of the Apostles, and the cloisters of the Prophets, the cloisters of the Martyrs, the cloisters of Saints, and the cloisters special for each rank of saints. Each monastery was distinguished by its extraordinary beauty, and in terms of length and width I could compare each with Constantinople if only they were not even better and did not have many bright rooms not made by hand. All those who were there, seeing me, rejoiced at my salvation, met and kissed me, glorifying God, who delivered me from the evil one.

When we went around these cloisters, I was sent down to the underworld, and there I saw the unbearably terrible torments that are prepared in hell for sinners. I heard screams and weeping and bitter sobs there; some groaned, and others angrily exclaimed: alas for us! There were those who cursed the day of their birth, but there was no one who would pity them.

Here, the analogy with the series of trials among the ancient Egyptians is quite noticeable, and less noticeable with the gates passed by Inanna and the fields passed by Gilgamesh.

As for the comparison with the general post-mortem scenario, the cited tests can be compared with stages of transition 1 and 3, stage 2 of choice (Table 4, Chapter 11), and at the stage of choice, a person's lifetime desires and actions play a decisive role.

Fortieth Day and Beyond

We quote (St. Macarius of Alexandria, 1831): on the fortieth day, the soul again ascends to worship God; and then the Judge determines a decent place for her in deeds. So, the Church is doing the right thing in making a commemoration of the departed and those who have received Baptism. This private judgment of God can be carried out quite routinely (The Ordeal of St. Theodora, n / d):

Having finished examining the places of torment, the angels took me out of there and brought me to the monastery of St. Basil, saying to me: "Now the Monk Basil is making a memory of you." Then I realized that I had come to this resting place forty days after my separation from the body.

This is the end of transition stage 3 and the beginning of certainty stage 4; the state of souls depends on where they are (for the Orthodox - in heaven or hell, as follows from the parable of the rich man and Lazarus (Luke 16:19-31), or in a special Purgatory, as among Catholics). But in torment or bliss, the soul retains its faculties (partial convolution). Her condition can only be changed by the prayers of the Church or personal ones. So, St. Gregory the Great, Dialogist, Pope of Rome (540-604), answering the question: *"Is there anything that could be useful to souls after death", teaches: that the holy sacrifice of Christ, our saving Sacrifice, is of great benefit souls even after death, provided that their sins can be forgiven in future life. Therefore, the souls of the departed sometimes ask that the Liturgy be served for them... Naturally, it is safer to do during our lifetime what we hope others will do about us after death. It is better to make the exodus free than to seek freedom in chains. Therefore, we should despise this world from the bottom of our hearts, as if its glory had already passed, and daily offer God the sacrifice of our tears as we offer His sacred Flesh and Blood. Only this sacrifice has the power to save the soul from eternal death, for it mysteriously represents to us the death of the Only Begotten Son"* (St. Gregory the Dialogist, 1996).

This stage, located to the right of the blue circle in Figure 12 of Chapter 11, can, by analogy with the state of the body after death described in the same chapter, be metaphorically called "mummification of the soul" - while retaining the ability to perceive and think, it enjoys limited freedom and can already act in the world of the living otherwise than by special grace.

Terrible Judgment

The events of the Last Judgment should be associated with the *decay of the world and the creation of everything new*, and not with the implementation of a posthumous scenario. St. John of Damascus (St. John of Damascus, 2011) describes the final state after death as follows:

We believe also in the resurrection of the dead. For there will be in truth, there will be, a resurrection of the dead, and by resurrection, we mean resurrection of bodies. For resurrection is the second state of that which has fallen. For the souls are immortal, and hence how can they rise again? For if they define death as the separation of soul and body, resurrection surely is the re-union of soul and body, and the second state of the living creature that has suffered dissolution and downfall. It is, then, this very body, which is corruptible and liable to dissolution, that will rise again incorruptible. For He, who made it at the beginning of the sand of the earth, does not lack the power to raise it up again after it has been dissolved again *and returned to the earth from which it was taken, in accordance with the reversal of the Creator`s judgment. ...*

Wherefore if it is the soul alone that engages in the contests of virtue, it is also the soul alone that will receive the crown. And if it were the soul alone that revels in pleasures, it would also be the soul alone that would be justly punished. But since the soul does not pursue either virtue or vice separate from the body, both together will obtain that which is their just due. ...

We shall therefore rise again, our souls being once more united with our bodies, now made incorruptible and having put off corruption, and we shall stand beside the awful judgment seat of Christ: and the devil and his demons and the man that is his, that is the Antichrist and the impious and the sinful, will be given over to everlasting fire not material fire like our fire, but such fire as God would know. But those who have done good will shine forth as the sun with the angels into life eternal, with our Lord Jesus Christ, ever seeing Him and being in His sight and deriving unceasing joy from Him, praising Him with the Father and the Holy Spirit throughout the limitless ages of ages. Amen. (St John Damascene, n / d).

The final post-mortem state of mankind allows for a direct comparison with the Egyptian counterparts - very few will receive new names and bodies *and follow Christ wherever he goes*; others who *at least once called on the Name of the Lord* will be saved, and, perhaps, they *will eat, drink, and have fun* in their Cane Fields, enjoying the eternal freebie; *and whoever was not written in the book of life was thrown into the lake of fire* - in these words of St. Ap. John the Theologian describes the fate of the dead as if quoting the Book of the Dead again and recalling the scenario of final death.

Only then will the stabilization scenario be completed. As we can see, the variant of the generalized posthumous scenario with the achievement of the merger is not considered - the creature cannot become equal with the Creator. And yet: *and the dust returned to the earth as it was, and the spirit returned unto God who gave it* (Eccl. 12:7) – this is already very reminiscent of the final phase of the fusion stage 5.

Let us summarize the scenarios of different epochs and cultures in Table 1, convenient for various comparisons.

Table 1. Comparison of the main post-mortem scenario with the high-ranking post-mortem sources; + means the stage description presence in the source

Sources	Stages of the main post-mortem scenario					
	0 - initial	1 - transition	2 - choice	3 - transition	4 - distinctness	5 - merge
Sumerian-Akkadian myths		+	+	+	+	
Egyptian Book of the Dead and Commentary	±	+	+	+	+	
Bible, St. Fathers, exegesis, Tradition	±	+	+	+	+	±

FUTURE RESEARCH DIRECTIONS

In the next chapter, the verification of the events of the posthumous scenario will be continued using another, independent, group of sources - the texts of the Indo-Tibetan region.

CONCLUSION

As follows from the reviewed texts and Table 4, Chapter 2, there is a good correspondence between the general post-mortem scenario and its particular cases and the content of the sources. We can conclude that they are all projections of the ternary connective. Having a common invariant (a general post-mortem scenario), the sources differ in calibrations associated with cultural and religious characteristics, and states of consciousness characteristic of different eras. Many texts make it possible to estimate the duration of events after biological death and up to the end of the general post-mortem scenario. This period is individual and usually lasts several weeks according to "Earthly" time. The moment of the end of the posthumous scenario and transitions to reincarnation or to merge with the light are also described in the sources; the interpretation of the transition of the wave into antiphase is unclear.

Thus, posthumous existence can be understood within the framework of the theory of complex systems self-organization as a continued evolution of open soliton-wave system components after its breakdown. There is no mysticism in this.

To verify the post-mortem scenario, in the absence of laboratory measurements, the only method available was used - a comparison with the universal human experience expressed in artifacts, symbols, rituals, and texts. And here, given the attention to the tendency of the ancient mind to personify objectively existing processes, such mysticism already appears.

Let's designate another interesting topic. In Chapter 2, an assumption was made about the duration of the main post-mortem scenario, based on the correlation with the flesh decay.

In ancient Egypt, the preservation of the physical body was given exceptional importance - without this, the preservation of personality traits and the success of posthumous wanderings was impossible.

In Christianity, there is a practice of veneration of relics, which in the popular mind is perceived as a sign of holiness. The Athos monks shared this view; The Church has never officially expressed such a position.

All this suggests that some correlation between the state of the remains and posthumous consciousness still exists and the assumption was correct.

But still, how reliable can the general post-mortem scenario and the materials supporting it be considered?

If the mathematical modeling methods are applied correctly, they are self-sufficient, and this is confirmed by the impossibility of the natural sciences' existence without mathematics.

Also, the cultural and religious heritage, different in details, but similar in meaning, when it comes to pre-mortal existence, are convincing for the humanities. Built on their basis following the method developed in Chapter 3, ternary connectives still lead to intuitive knowledge.

REFERENCES

Coleman, G. (Ed.). Jinpa, T. (Ed.), Dorje, G. (Transl.), Dalai Lama (Comment.). (2007). *The Tibetan Book of the Dead: First Complete Translation.* Penguin Classics

Collins, A. (2014). *Göbekli Tepe: The Origin of the Gods.* Bear & Company.

Curry, A. (2020). DNA from ancient Irish tomb Realitys incest and an elite class that ruled early farmers. *Science.* Advance online publication. DOI: 10.1126/science.abd3676

Dedović, B. (2020). *"Inanna's Descent to the Netherworld": A centennial survey of scholarship, artifacts, and translations.* Digital Repository at the University of Maryland. https://doi.org/DOI: 10.13016/ur74-yqly

Foster, B. R. (2001). *The Epic of Gilgamesh.* W.W. Norton & Company.

George, A. (2014). *Epic of Gilgamesh.* Penguin Classics.

Goelet Jr, O., Faulkner, R. O., Andrew, C. A. R., Gunther, J. D., & Wasserman, J. (2015). The Egyptian Book of the Dead: The Book of Going Forth by Day: The Complete Papyrus of Ani Featuring Integrated Text and Full-Color Images.

Holy Bible - American Standard Version. (2019, April 1). Holy Bible - American Standard Version - ASV. https://holy-bible.online/asv.php?

Kovalyov, Y., Mkhitaryan, N., & Nitsyn, A. (2020). *Self-organization of the human mind and the transition from paleolithic to behavioral modernity.* IGI Global

Johnston, S. A. (2006). *Inside the Neolithic Mind: Consciousness.* Cosmos, and the Realm of the.

Lascaux cave. Free placement of drawings and use of the relief of the walls of the cave. [Illustration]. Lascaux cave, France. (n.d.). Yandex.ru. Retrieved August 15, 2024, from https://yandex.ru/collections/card/5a07c8eb0c1ed2002fed3a3e/

Mercer, S. A. B. (Ed.). (1952). *The Pyramid Texts.* (n.d.). Sacred-texts.com. Retrieved August 15, 2024, from https://www.sacred-texts.com/egy/pyt/index.htm

Newgrange Tomb, Newgrange, Ireland. Approximately 3200 years. BC. [Illustration]. [Used Files]:

Newgrange Tomb. Modern view after restoration. (n.d.). [Newgrange. 2 February 2018]. Travellan.ru. Retrieved August 15, 2024, from https://travellan.ru/articles/nyugreyndzh/

Newgrange Tomb. The design of the mound, corridor, and chamber. (N.d.). Rumpus. Ru. Retrieved August 15, 2024, from https://rumpus.ru/turizm/nyugrejnzh-kurgan-fej-v-irlandii/

Newgrange Tomb. The path of the sun's rays on the winter solstice day. [Rick Doble]. *Computing the Winter Solstice at Newgrange: Comparing Neolithic science to Greek or Roman science*. (n.d.). Newgrange.com. Retrieved August 15, 2024, from https://www.newgrange.com/winter-solstice-newgrange.htm

Newgrange Tomb. Sun rays in the corridor on the winter solstice day. [Newgrange Winter Solstice. December 19, 2012]. *Irish History*. (n.d.). Blogspot.com. Retrieved August 15, 2024, from http://history-ireland.blogspot.com/2012/12/newgrange-winter-solstice.html

Newgrange Tomb. View of the central chamber. [Newgrange. 2 February 2018]. (n.d.). Travellan.ru. Retrieved August 15, 2024, from https://travellan.ru/articles/nyugreyndzh/

Newgrange Tomb. A stone with images of spirals at the entrance. [Rick Doble]. *Computing the Winter Solstice at Newgrange: Comparing Neolithic science to Greek or Roman science*. (n.d.). Newgrange.com. Retrieved August 15, 2024, from https://www.newgrange.com/winter-solstice-newgrange.htm

Old European culture. (n.d.). Blogspot.com. Retrieved August 15, 2024, from https://oldeuropeanculture.blogspot.com/2016/12/newgrange.html

Schmidt, K. (2020). *Sie bauten die ersten Tempel: Das rätselhafte Heiligtum am Göbekli Tepe*. C.H. Beck.

St. John Chrysostom. Archbishop of Constantinople. *The Paschal Sermon*. (n.d.). Oca.org. Retrieved August 15, 2024, from https://www.oca.org/fs/sermons/the-paschal-sermon

St. John of Damascus. (2008). *The funeral hymns of St. John of Damascus*. (n.d.). Blogspot.com. Retrieved August 15, 2024, from http://full-of-grace-and-truth.blogspot.com/2008/12/funeral-troparia-of-st-john-of-damascus.html

Tann. (2019). A detail from the floor of a coffin of Gua, physician of Djehutyhotep, a nomarch of Deir el-Bersha, Egypt, during the Middle Kingdom. [Illustration]. Source: Remains of 4000-year-old Egyptian Guide. Archaeonewsnet Date: 2019 Author: Tann. https://archaeonewsnet.com/2019/12/remains-of-4000-year-old-egyptian-guide.html

The Ordeal of St. Theodora. (N.d). Archaeonewsnet.com. Retrieved August 15, 2024, from https://katolyki-krasnodara.ru/en/mytarstva-prepodobnoi-feodory-mytarstva-blazhennoi-feodory-chto.html

ADDITIONAL READING

Bader, O. N. (1967). *Pogrebeniya v verhnem paleolite i mogila na stoyanke Sungir.* [Burials in the Upper Paleolithic and a grave at the Sungir site]. *Soviet Archeology 1967*(3). https://arheologija.ru/bader-pogrebeniya-v-verhnem-paleolite-i-mogila-na-stoyanke-sungir/

Danylenko, V. M. (1986). *Kamyana Mohyla* [Stone tomb]. Naukova Dumka.

Dergachev, V. A., & Manzura, I. V. (1991). *Poghrebaljnye kompleksy pozdnegho Trypoljja* [Burial complexes of late Trypillia]. Shtiintsa.

Dyakonov, I. M. (2006). *Epos o Ghyljghameshe (O vse povydavshem)* [The Epic of Gilgamesh ("About who has seen everything")]. Nauka.

Granin, R. S. (2019). Eskhatologhycheskaja paradyghma y ee struktura. [Eschatological paradigm and its structure]. *Interdisciplinary Research*, 2019, 133–147.

Ig. Hilarion (Alfeev). (2001). *Prep. Symeon Novыj Boghoslov y pravoslavnoe predanye*. [Rev. Simeon the New Theologian and Orthodox Tradition]. Alethea.

Human triparties. (n.d.). Azbyka.ru. Retrieved August 15, 2024, from https://azbyka.ru/otechnik/prochee/dusha-chelovecheskaja-polozhitelnoe-uchenie-pravoslavnoj-tserkvi-i-svjatyh-ottsov/19

Ivan Chelovekov. (n / d). Typology of Dolmens. [Illustration]. Retrieved August 15, 2024, from https://history.eco/dolmeny-polumonolity/

Khazarzar, R. Library. (n / d). Retrieved August 15, 2024, from http://khazarzar.skeptik.net/books/shumer/inanna.htm

Kindness (1992). (5 vol.). Holy Trinity Sergius Lavra.

Korvin-Piotrovsky, A. G. (2008). *Trypiljsjka kuljtura na terytoriji Ukrajiny* [Trypillia culture on the territory of Ukraine]. Institute of Archeology of the National Academy of Sciences of Ukraine.

Ksenofontov, G. V. (1992). *Shamanizm*. [Shamanism. Selected works.]. North-South. (Original work published 1928-1929).

Kuzmin, Y. V. (2023, October 9). *Prjamoe radyoughlerodnoe datyrovanye parnogho poghrebenyja na stojanke verkhnegho paleolyta Sunghyrj: novye rezuljtаты, starye problemy.* [Direct radiocarbon dating of a paired burial at the Upper Paleolithic site of Sungir: new results, old problems]. Antropogenez.ru. https://antropogenez.ru/article/473/

St. Luka (Voyno-Yasenetsky). (2013). *Duh, dusha i telo*. [Spirit, soul, and body]. Dar.

Misteria Stounxdza. (2008, May 31). [The Stonehenge Mystery]. RBC. https://www.rbc.ru/society/31/05/2008/5703cccf9a79470eaf76ae69

Mykhailov, B. D. (2005). *Petroghlify Kam'janoji Moghyly: Semantyka. Khronologhija. Interpretacija* [Petroglyphs of Kamyana Mohyla: Semantics. Chronology. Interpretation]. MAUP.

Mytarstva sv, Teodory. [Ordeals of St. Theodora]. Pravmir.ru. Retrieved August 15, 2024, from https://lib.pravmir.ru/library/readbook/1959

Newgrange. 2 February 2018. Travellan.ru. Retrieved August 15, 2024, from https://travellan.ru/articles/nyugreyndzh/

Newgrange: Fairy Mound in Ireland. (N.d.). Rumpus.Ru. Retrieved August 15, 2024, from https://rumpus.ru/turizm/nyugrejnzh-kurgan-fej-v-irlandii/

Pminoveniya usopshih. (2016). [Prayers for the dead]. Palomnik.

Primitive Sungir. (n / d). Paired burial of adolescents at the Sungir site. [Illustration]. Retrieved August 15, 2024, from https://musei-smerti.ru/pochemu-pervobyitnoe-zahoronenie-sungir-tak-unikalno/

Rak I. V. (1993). *Mify Drevnego Egipta*. [Myths of Ancient Egypt]. Publishing house "Petro - RIF".

Shaposhnikov, A. K. (Ed.). (2003). Drevneeghypetskaja Knygha Mertvykh. Slovo ustremlennogho k Svetu. The Ancient Egyptian Book of the Dead. The word of one striving towards the Light. Eksmo Publishing LLC

Smyrnov, V. (n.d.). *Stounkhendzh v fotoghrafyjakh XIX y XX veka*. [Stonehenge in 19th and 20th-century photographs]. [Illustration]. Mediasole.Ru. Retrieved August 15, 2024, from https://interesnosti.mediasole.ru/stounhendzh_v_fotografiyah_xix_i_xx_veka

St. Gregory Nyssky. (1862). *Tochnaja ynterpretacyja Pesny Solomona*. [The exact interpretation of the Song of Solomon]. In *Creations of the Holy Fathers in Russian translation*. v. 39: Works of St. Gregory of Nyssa. part 3, 298

St. Gregory Palamas. (2004). *Vsechestnoj v monakhyne Ksenyy, o strastjakh y dobrodeteljakh y o plodakh umnogo delania.* [To the all-honorable in the nun's Xenia, about passions and virtues and about the fruits of smart doing] In *Philokalia* (Vol. 5). Sretensky Monastery Publishing House.

St. Gregory Palamas. (2011). *Tryady v zashhytu svjashhennbezmolvstvuyuschyh.* [Triads in defense of the sacred silent ones]. Academic project.

St. Grigory Dvoeslov. (1996). *Voprosy o zhyzny ytaljjanskykh otcov y bessmertyy dushy* [Questions about the Italian Fathers' life and the soul immortality]. Blagovest.

St. John Damascene. (n / d). *Tochnoe izlozenie pravoslavnoj very.* [An Exact Exposition of the Orthodox Faith].

St. Macarius of Alexandria. (1831). Slovo ob yskhode dush pravednykh y ghreshnykh. [A word about the outcome of the souls of the righteous and sinners]. *Christian Reading*, (part 43), 123–131.

St. Simeon the New Theologian. (2011). *Tvorenyja prepodobnogho Symeona Novogho Boghoslova. Slova y ghymny.* Works of St. Simeon the New Theologian. Words and hymns. (3 books). Sibirskaya Blagozvonnitsa.

Vasiliadis, N. (2012). *Taynstvo smerti.* [Sacrament of death]. Publishing House of the Holy Trinity Sergius Lavra. https://pravoslavnoe.uaprom.net/p782708550-vasiliadis-tainstvo-smerti.html

ENDNOTES

[1] Some researchers dispute the fact of burial, believing that the bones were brought by animals, or a sacrifice took place.

[2] It is amazing how close the gods are to people in myths, differing from them, however, in the power of the word, the ability to determine fate and immortality.

[3] The plot played out by Ursula Le Guin in A Wizard of Earthsea is known from many myths.

[4] In the Upanishads, the following legend is stated: the souls of the righteous go to the Sun and no longer incarnate; the rest of the souls go to the moon and after the full moon they return with rains to the earth for new incarnations. The Egyptians are not unique in the idea of "the path of the sun".

⁵ All of the above is not recognized by authoritative Christians of the Protestant confessions.

Chapter 6
Events Checking:
Indo-Tibetan Version

ABSTRACT

The verification of the main posthumous scenario invariants – events and transitions – (Chapter 2) showed their correspondence with the data of the Indo-Tibetan region sources. Some differences between the texts of the Upanishads and the Tibetan Book of the Dead are associated with changes in human consciousness from era to era. This leads to a change in the initial data for the process of folding and modifying the main scenario. Calibrations allow us to find out how consciousness perceives the invariants of the general posthumous scenario, and what expressive means the authors of the texts use.

BACKGROUND

Sources were used by their ranks, supplementing them as needed. We will pay more attention to data about the post-mortem events.

Before moving on to a detailed examination of post-mortem events in the Tibetan Book of the Dead, let us compare it with an earlier Indian source - the Upanishads (Swami Nikhilananda, 1986-1994). Such a comparison will help us better understand the context of the Tibetan book and see the most serious differences in the death phenomenon interpretations.

The Tibetan Book of the Dead gives many details (Bardo Thodol, n / d; Tibetan Book of the Dead, 1992; Coleman, 2007; Thurman, 2011).

All sources will be used to verify post-mortem scenarios, including invariants and calibrations; the sources will be compared with post-mortem scenarios (Chapter 2) and each other.

DOI: 10.4018/979-8-3693-9364-2.ch006

THE TIBETAN WAY

We need to start with the creation myth. So, according to the Aitereya Upanishad (Ch. 1, Sect. 1):

1. In the beginning all this verily was Atman (Absolute Self) only, one and without a second. There was nothing else that winked. He (Atman) willed Himself: "Let Me now create the worlds".
2. He created these worlds: Ambhah, the world of water-bearing clouds, Marichi, the world of the solar rays, Mara, the world of mortals and Ap, the world of waters. Yon is Ambhah, above heaven; heaven is its support. The Marichis are the interspace. Mara is the earth. What is underneath is Ap.
3. He bethought Himself: "Here now are the worlds. Let Me now create world-guardians." Right from the waters He drew forth the Person in the form of a lump and gave Him a shape.
4. He brooded over Him. From Him, so brooded over, the mouth was separated, as with an egg; from the mouth, the organ of speech; from speech, fire, the controlling deity of the organ. Then the nostrils were separated; from the nostrils, the organ of breath; from breath, air, the controlling deity of the organ. Then the eyes were separated; from the eyes, the organ of sight; from sight, the sun, the controlling deity of the organ. Then the ears were separated; from the ears, the organ of hearing; from hearing, the quarters of space, the controlling deity of the organ. Then the skin was separated; from the skin, hairs, and the organ of touch; from the hairs, plants trees, air the controlling deity of the organs. Then the heart was separated; from the heart, the organ of the mind; from the mind, the moon, the controlling deity of the organ. Then the navel was separated; from the navel, the organ of the apana; from the apana, Death, Varuna, the controlling deity of the organ. Then the virile member was separated; from the virile member, semen, the organ of generation; from the semen, the waters, the controlling deity of the organ.

Ch. 1 – Sect. 2:

1. These deities, thus created, fell into this great ocean. He, the Creator, subjected the Person (Virat in the form of a lump) to hunger and thirst. They (the deities) said to Him (the Creator): "Find out for us an abode wherein being established we may eat food."
2. He brought them a cow. They said: "But this is not enough for us." He brought them a horse. They said: "This, too, is not enough for us." He brought them a person. The deities said: "Ah, this is well done, indeed." Therefore, a person is

verily something well done. He said to the deities: "Now enter your respective abodes."

3. The deity fire became the organ of speech and entered the mouth. The air became breathable and entered the nostrils. The sun became sight and entered the eyes; the quarters of space became hearing and entered the ears. Plants and trees, the deity of air, became hairs and entered the skin. The moon became the mind and entered the heart. Death became the apana and entered the navel. The waters became semen and entered the virile member.

Thus, at a certain stage of creation, Atman sacrificed their personality (Purusha) to the gods, i.e., in a certain sense, himself. This sacrifice established balance in the world, nourishing the gods and all that exists and becoming the basis for the relationship between the parts of the human body and the various parts of the world[1].

This is what Yama, the first man to die, did, turning his death into an act of sacrifice and becoming the embodiment of death, a metapomp, and the ruler of the world of the dead, and other people do the same.

Thus, death sacrifice is the most important act of paying a debt to the gods, maintaining the world order, and choosing the path of one's posthumous existence.

According to the Brihadaranyaka Upanishad (Part 6, Ch. 2):

13. "Woman, O Gautama, is the fire, her sexual organ is the fuel, the hairs the smoke, the vulva the flame, sexual intercourse the cinders, enjoyment the sparks. In this fire the gods offer semen as libation. Out of this offering a man is born. He lives as long as he is to live. Then, when he dies,
14. "They carry him to be offered in the fire. The fire becomes his fire, the fuel his fuel, the smoke his smoke, the flame his flame, the cinders his cinders, and the sparks his sparks. In this fire, the gods offer the man a libation. Out of this offering, the man emerges in radiant splendor.
15. "Those even among householders who know this, as described and those too who, living in the forest, meditate with faith upon the Satya Brahman (Hiranyagarbha), reach the deity identified with flame, from him the deity of the day, the deity of) the fortnight in which the moon waxes, from him the deities of the six months during which the sun travels northward, from them the deity identified with the world of the gods (devaloka), from him the sun, from the sun the deity of lightning. Then a being created from the mind of Hiranyagarbha comes and leads them to the worlds of Brahmin. In those worlds of Brahma, they become exalted and live for many years. They do not return to this world again.
16. "But those who conquer the worlds through sacrifices, charity, and austerity reach the deity of smoke, from smoke, the deity of the night, from night the deity of the fortnight in which the moon wanes, from the decreasing half of

the moon the deities of the six months during which the sun travels southward, from these months the deity of the world of the Manes and from the world of the Manes, the moon. Reaching the moon, they become food. There the gods enjoy them, just as here the priests drink the shining soma juice-saying as it were: "Flourish, dwindle." And when their past work is exhausted, they reach this very akasa, from the akasa they reach the air, from the air rain, from rain the earth. Reaching the earth, they become food. Then they are again offered in the fire of man and thence in the fire of woman. Out of the fire of woman they are born and perform rites with a view to going to other worlds. Thus, do they rotate? "Those, however, who do not know these two ways become insects and moths and those creatures which often bite (i.e. mosquitoes and gnats)."

The Katha Upanishad, written in the form of a dialogue between Yama and Nachiketas, talks about how to avoid the eternal cycle of births and deaths (samsara) - one must, as it were, go through the path of creation in the opposite direction, identifying one's individual atman with the world Atman, devoid of any duality, including death/immortality.

According to the Katha Upanishad (Ch. 3):

1. In order to reap the results of one's own actions done in this world, two are said to be entering the cave of heart and residing there. The wise men, the worshippers of the five-fold aspect of Agni and those who have completed the Naciketa-agni worship atleast thrice – compare these two to light and shadow (the actual comparable simile is that of Jivatma and the Paramatma).
2. I think this practice of Naciketa-Agni shall serve like a bridge (unto the Supreme) for those who perform Yagnas or sacrifices, and like the immortal Brahman and shall be the place of no fear for those who want to cross the ocean of births and deaths.
3. If the body is the chariot, the mind is the steering cord (of the horse), the intellect is the driver, and the traveler is the Jiva or the Purusa. Understand it to be so.
4. The horses (driving the above chariot) are the sense organs and the materialistic desires or things of interest are the paths they travel by. The wise ones call the Jiva, which comprises of the body, the mind and the sense organs as 'Bhoktha'.
5. In as much the same way that the wild horses cannot be tamed by a rider, the sense organs will be uncontrollable by that one who has no control over his mind and has no sense of intellect.
6. (On the other hand), the sense organs will be fully under the control of that one who has complete control over his mind and has the distinguishing power of intellect. This is comparable to the control a rider exercises over trained, tamed, good horses.

7. That one who has no intellect, who has no control over his mind and who is dirty (in his mind and thoughts), he will not reach the destination (of the Supreme). He will continue to wander in this world through further births.

8. (On the other hand), that one who is full of intelligence and who has complete control over his mind and is very clean (in his mind and thoughts), he reaches that place from where nothing is ever born again.
9. That one who has the intelligence as the rider and a controlled mind as the steering cord, he travels the long way and ultimately reaches the end of the tedious path of samsaara (or the materialistic world). That is the state (called as the Parama-pada) where there is a unison between the Jiva and the all-pervading Divinity.
10. Beyond the sense organs are the objects of sense. Beyond them is the mind. Beyond it is the brain or intuition. Beyond that is the great soul (one of the aspects of the Atma).
11. Beyond that great soul is the unmanifest divine. Beyond that is the Purusha (the all-knowing and all-pervading Atma). There is nothing beyond the Purusha. That is the end. That is the supreme.
12. This purusha remains hidden in all beings and creatures; he is not visible (to the naked eye). He is seen only by those wise seers who are able to do so by concentrated efforts and focused meditation.
13. The wise man with subtle intelligence shall restrain the speech in his mind, the mind in the intellect, the intellect in the great soul, and the great soul in the Supreme Divine (from where there is no return to this materialistic world).
14. Arise! Awake! Attain the pure knowledge by following the intelligent teachers who practice what they preach. The experienced sages say that the path (leading to self-knowledge) is a risky one since it is sharp as the edge of a knife and also difficult to tread upon or cross over.
15. Having known that which is beyond the aspects of sound, feel, vision, taste and smell, which is unchanged and eternal, which has neither a beginning nor an end, which is beyond even the greatest, which is stable and consistent, one frees himself from the mouth of death.

Note that all this corresponds to yogic meditation: at first the senses are removed from the objects of perception, then the internal dialogue ceases, then the distinction of opposites, after which the feeling of one's ego gradually fades, which leads to the state of samadhi.

For followers of the Buddha, who from the very beginning deny sacrifices as such, the Vedic understanding of death is deprived of its sacredness, the cessation of samsara is associated with deliverance from suffering, and the achievement of this goal turns into an applied task, solved through enhanced meditation practice.

This is the main difference between Buddhism and the Vedas, Upanishads, and classical darshans. From these positions, the text of the Tibetan Book of the Dead was written, thanks to which we have many details of the posthumous state that are not in the Upanishads.

Now let's return to the Tibetan Book of the Dead. The text written specifically for reading over the body of the deceased and calculated that the deceased would hear it is written with the utmost clarity. Therefore, it can simply be quoted, and there is no need to use comments to interpret it. Several translations are available to authors (Tibetan Book of the Dead, 1992; Coleman, 2007; Thurman, 2011). We will quote (Bardo Todol, n / d), choosing this source because of its brevity and clarity of presentation, despite the inaccuracies of the translation. For convenience, let's preface the presentation of the text with Figure 1, showing the sequence of major events.

Figure 1. Post-mortem events according to (Bardo Todol, n / d)

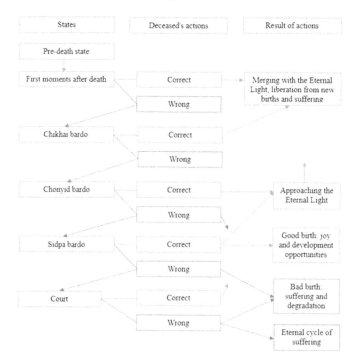

The states and actions in more detail, accompanied by comments and comparisons with the post-mortem scenarios described in Chapter 2.

Pre-Death State

The time of your departure from this Reality is approaching. The signs of Death in sensations are as follows: the Earth is immersed in Cold Water. The burden is filled, plunging, with cold. Chilliness and lead engorgement; Water turns into Fire. He throws it into the heat, then into the cold; Fire turns into Air. Explosion and Disintegration by fading sparks in the void. These elements prepare us for the moment of death, mutually changing. When Fire scatters in the Void of Air, it means the time has come for you to enter the Chikhai bardo space.

The agonal state is described in terms of the 5 elements.

The First Moments After Death

Soon you will exhale with your last breath, and it will stop. Here you will see the eternal Pure Light. Unbelievable Space will open before you, boundless, like the Ocean without waves, under a cloudless sky.
Like a fluff, you will float, freely, alone.
Do not be distracted, do not rejoice! Don't be afraid! This is the moment of your death! Use death, for it is a great opportunity. Keep your thoughts clear, not clouding them even with compassion. Let your love become passionless.
If you see the Glitter, it is the Glitter of the Eternal Light of the Enlightened Reality. Understand that. Your present Consciousness, not filled with impressions, sounds, pictures, or smells, perceives Itself, which is the real Reality.
Your mind, no longer non-existent, gaping with eternity, is not emptiness or unconsciousness. Left only to Yourself, it sparkles, flashes, burns - this is your real purified Consciousness.
Your consciousness and your sparkling mind are inseparable, they are the same. Their union is Dharma-Kaya, the state of Perfect Illumination by the Eternal Light.
You are now aware of the brilliance of your own purified, nonexistent Mind.
It is enough to understand only this. Having recognized that in the yawning of the eternity of the mind, there is a frighteningly bright Enlightenment, perceiving it at the same time as one's consciousness, this means to keep oneself in the (state of) divine enlightenment of the Buddha.

Your Consciousness, gaping, dissolved, and inseparable from the Great Shine of Eternity, has no birth and does not know death. It is the Eternal Light - Buddha Amitabha.

Collect yourself, and look with your eyes for the Eternal Light. When you see it, take it! Here he is! Don't let your attention wander. This is a direct encounter with the ultimate truth, the Law (Dharma-Kaya). You will be able to see, and recognize - you will become what you are. You will know the secret, you will know the haze of Life and Death, and you will become this Light yourself. This is a vertical path accessible to a few.

As northern Buddhism teaches, using the "Great Direct Vertical Path" one can immediately free oneself, even achieve Bliss, without stepping into the Bardo spaces at all, without being burdened by the long road of the usual path crossing countless plains and gorges of karmic mirages. This possibility underlies the whole essence of the Bardo Teaching as a whole. Faith is the first step on the Secret Path. Following the first step, the second step is Illumination. Only Illumination insures us against losing the ground under our feet on the true path. With Illumination comes Certainty. And when the Exhaustion of Aspiration is reached, then Freedom comes. Success on this Secret Path depends entirely on the development of the soul. If a mortal is capable of embracing what is Reality to him, if he can die in consciousness, with full consciousness of his soul at the moment when he parted with his body if he shouts to the descending terrible light: "You are I," then all the chains of samsara will burst, and the Sleeper will wake up in the Only Reality.

Simeon the New Theologian would probably subscribe to such a description of the Light, but would also see a difference: in Christianity, a person can do god-like deeds, even be called a god (*God standeth in the congregation of God; He judges among the gods. Ps. 82:1*), but God is separated from the creature by the fact of its creation. In Buddhism, the soul is inseparable from the universal spirit (Tat tvam asi - You are That), the ultimate goal is precisely the dissolution of the personality in the universal spirit, and this is the only way to avoid endless suffering in the cycles of samsara.

Therefore, a general post-mortem scenario can be implemented here, and the quote describes exactly how to do this simply - to implement the convolution scenario immediately, in one transition, without intermediate stages. Naturally, to be able to do this in the afterlife, one must work hard here, in the present existence.

Chikhai Bardo

You did not see the Eternal Clear Light. In anticipation of the next Bardo, Secondary Clarity may light up in front of you.

The time of the secondary Light lasts several hours after the breath has stopped. The vital force leaves the body through one of the openings, and here comes the clarification. Once out of the body, the first thing Consciousness asks is - am I dead or not? We see relatives, friends, or doctors, as we are used to it, we even hear what they are talking about. "Where am I? Our Essence, which has awakened, asks, "If my body lies there?" We soar within the same limits of places of employment, and people, in life. Looking around ourselves and focusing on the details of a hand or palm, for example, we will find that we have become transparent, that our new body is just a play of light, glare. It is worth recognizing this and not being afraid - Salvation will come in an instant. The Secret Path will open.

This is the Secondary Light of Eternity. In reality, having learned something new in ourselves and accepting it, we change, we become different. So it is with the Light, having recognized ourselves in it, we become it! Without recognizing, without seeing the light, here we can meet the Sentinels of Eternity in any guise. If you did not see the light, but you woke up, came to your senses, and know where you are, keep one thought in your head: humility! Whomever you meet, bow down and remember at least some prayer from a past life, at least something that you believed in. The appearance of Sentinels usually corresponds to our life habits. A dying child may have a visit from his parents, mother, or father. Submit and rejoice in the guide. Turn to him with a prayer and the Great Path will open.

Chikhai bardo lasts 3–3.5 days.

For many, this is a time of unconsciousness. During this unconsciousness, the Eternal Light illuminated you, but you were like a body under the Sun, lying unconscious on the clay.

And without waking up, you lay under the Eternal Light sparkling from everywhere in Chikhai bardo. Now you will emerge as a Consciousness, separated from the flesh that is no longer needed, into the Chikhai bardo. Be careful and attentive! Do not rush! Do not be afraid! You died! Understand this and do not cling to the departed, do not stir up feelings, do not let them run wild and swallow you. Waves of experiences can carry us to scary places. Pull yourself together and look around with an attentive and kind look. The Light will hang in front of you, like a bright mirage, playing and blinding. Inside the Light, you will hear thunders. These are the sounds of the last Essence. Do not be afraid! Nothing can hurt you, because you - no! So you can be whatever you want. Become that Sound, and respond to it. These mirages are you! The one who is not contains nothing and everything! If you don't recognize, don't respond to visions and sounds, if you don't see yourself and your own in them, fear will seize you! Through the broken ice, you will fall into other worlds of real misfortune and torment. Beware of involuntary feelings. Let the kindness of the hardness of pure, mirror glass be your main feeling.

The Bardo of Death lasts an average of 49 days, starting from the day the deceased realized their death. Usually in 3 days, this awareness comes. After that, until about 15 days, we wander to Chonyid Bardo. Then - the Judgment and Sidpa Bardo, in which we are looking for the coming, as a rule, forgetful birth[2]!

So, the sensations of the initial stage 0 are described (Table 4, Chapter 2). As in near-death experiences, only a small percentage of the deceased experience feelings of euphoria heightened feelings, and seeing themselves as light, and most do not feel anything.

The first can realize that the light of their consciousness is the light of the universal spirit and immediately end the post-mortem scenario or, conversely, begin to go through it in stages, asking the Sentinels of Eternity to become metapomps, starting transition stage 1 "with a guide".

For others, Chikhai bardo is a time of unconsciousness, but for them, too, it ends with transition stage 1, which is characterized by the weakening of all sensations (Table 4, Chapter 2).

Chonyid Bardo

All the days of this stage are described; to understand the essence of what is happening, it is enough to describe the first two.

The first day. *For three days you did not know what happened to you. Now, when you wake up, do not catch yourself - you have changed a lot. Everything has changed and you have become different. Do not look in the mirror and do not touch your loved ones. You will see things that you have not seen; You will hear sounds that are not like Earth. The source of both is the Circle (mandala) of your heart. This is you, in that inseparability of yours, where Illumination is available to us. The middle of the first day is the domain of the White Buddha, Vairochana. It has a blinding white body that glows with pure blue light. So bright is this Light, so terrible is its flame, that it is easy to be afraid. Whoever is not afraid, believes in the blue flame, and comes to his senses - will be saved from the great pain and torment of the Bardo.*

The pure blue flame is mixed with the calm white light of the Divas, the Demigods. The Book of the Dead advises us to avoid the white steady flame, it can carry away and hinder our path. The ways of the gods are different from the ways of men.

See Yourself in the White Figure shrouded in blue flame, and become the White Buddha!

The Second day. *This is the day of the clarity of the white fire of the Eastern side, in this white pure flame are the Happiness of Penetration and the Wisdom of the Mirror. On the same day, hell will appear for the sinner, dissolving its terrible mouth, from where dark light flows. Evil deeds or anger can push you, pull you irresistibly towards the smoky dark light of Hell. It will seem so warm, warm. The hard brilliance of salvation will terrify. Do not look in that seemingly gentle smoky dark side. This is the way to the hellish worlds, from where the way out will be long. Beware of anger, especially here in the Bardo. On this second day, you can still see those left behind, in the earthly vale, you can hear how they argue, dividing your property. You will understand that your beloved wife has forgotten you. God forbid, you will get angry - in an instant, the dark light will pull towards itself and the hellish door will open.*

A clear, pure white flame sparkles so brightly, so blinding that it hurts the eyes to look at it. Clear white fire is mixed with smoky black light, and Hell and Evil glow with this agate color. The evil in a person will reject the blinding white flame as someone else's, and the person will be tempted and follow the smoky, measured fire. Resist temptation: smoky black fire leads to suffering, to an uncertain and defenseless Future. Gaze into the bright shining white flame and take it into Yourself.

And so on - the deceased is presented with more and more terrible visions. In each case, behind the appearance, he must recognize the essence and make the right choice, which determines either the continuation of following the path or getting into certain places of suffering. This is an analog of the three series of trials of Ancient Egypt and, to some extent an analog of the ordeals in Christianity. The difference is, that by right choice making on the first or second day, you can avoid the rest of the tests. We refer to Table 4 of Chapter 11 - the stage of choice 2 is described.

The deceased, who missed the possibility of complete convolution immediately, but made the right choice of a living symbol, created a projection for himself and organized an incarnation in the new world (reincarnation scenario), using means similar to the magic of the ancient Egyptians, or fell into hellish worlds if the choice was wrong. However, not everyone was able to choose at least something. These move to the Sidpa Bardo.

Sidpa Bardo

It also describes the events by day, up to the new birth. Here is the most informative description of the fifteenth day, and several other interesting fragments.

Fifteenth day. *Chonyid Bardo flashed before my eyes like a dream, degenerating into a nightmare towards the end. You did not recognize Yourself in any circle (mandala), you ran, as if from terrible dreams, from the faces and figures that surrounded you: although how can our own dream harm us?!*

You are now in the Sidpa Bardo. Being born in the Bardo is not like being born on earth. Your consciousness emerges like a head from muddy water, and in an instant, you are there! Like a trout jumping out of the water, suddenly you wake up.

An eloquent description of transition stage 3, characterized by muted feelings. The state in the Sidpa Bardo is described as follows:

Your body is the same as before. All feelings are with you, you are free in motion. Possessing supernatural karmic powers, you can see other beings of the Bardo, and they see you. Your body is similar to your former one, but it is more perfect, because it was born from hopes and desires, signs of the future. If you look very closely, you will see a transparent play of lights instead of flesh. Get a little distracted - and again the flesh will harden before your eyes.

A very interesting fragment in which one can see the influence of the observer on the state of the system is an analog of the uncertainty principle.

Visions will visit you: your future birth, places, and those with whom you will live again. Do not follow them, do not deceive yourself - this is a long path of suffering. Here and now, it is extremely important not to lose yourself, not to be distracted, and not to let feelings captivate you. If you do not give in to the temptation at all, you will be freed without having to re-enter your mother's womb. You cannot keep your consciousness, you lose yourself, then think about your God, about the Teacher or the Person who shone in your life and warmed you.

Try to listen to this Knowledge. There is great skill in the Sidpa Bardo needed. In the sharpness of the senses, with the ability to move instantly at will, you can choke with joy. Hold on.

You are no longer constrained by the flesh, in this karmic body, you can penetrate the thickness of walls, rocks, and mountains. This is proof that you are in the Sidpa Bardo. The supernatural forces of karma will allow you to find yourself in a chosen place in an instant. You can change your size, and shape, and even exist in several faces[3]. Do not be frightened or deceived by these new abilities.

Note: the absence of space and time (instantaneity of movement), as well as the usual physical properties - a consequence of the weakening of the practical mind (level 4); the possibility of complete convolution in the course of a generalized post-mortem scenario remains as a potential one, but it can be realized only by taking control of the will and reason.

Finally, you will see people and hear voices in the Earthly Vale. You will speak to them, but they will not hear you. If you touch them, they will pass through you without noticing. They cannot see or hear you. To look at native faces point-blank and not be noticed; to hear the voices of loved ones and not be able to call out to them - what a terrible grief the soul can plunge into! Take comfort in thinking about the Lord and the Teacher, if you had one. At all times, a silvery light will accompany you - this is the illumination of the Sidpa Bardo.

From one week to seven, until the 49th day, you will stay in this state. Most remain in this state for 22 days; the exact term is determined by Personal Karma.

The wind of Karma will push you in the back, but not a single branch near you will move. This is the wind of your Karma; it pushes you because it originates in You! Don't be afraid! Darkness and Blackness will spread ahead, from where frightening cries piercing the soul will be heard. These are your fears - their beginning is in you! Bad Karma can hit you with demons, wild beasts will chase you. A terrible storm will break out on the way, and an angry crowd will rush towards you, threatening to trample, tear to shreds. This evil originates in ourselves. It is our own nightmares and biting conscience. If you are not afraid, they are powerless, its threats are unrealistic. Go towards and hug, kiss the bloody mouth - and the visions will disappear!

If you get scared and run, you will find yourself in a gorge, on a terrible cliff, from where there is no way. Three abysses will cut off your path: a white abyss, a black one, and a red one. These are the three Evils: Anger, Selfishness, and Stupidity. If you see these three abysses, know that you are in the Sidpa Bardo. Focus on yourself, turn to the Lord, and strengthen yourself - nothing and no one can harm you here, but you have to believe in it.

Who during life tried to see Himself - happiness and joy will blow on them, wonderful views will open for them, smells will flow, caressing the nostrils, and the light maidens will come with a gentle step. Don't be tempted! This also comes from you, do not deceive yourself.

He who lived as in emptiness will see meaningless paths, emptiness, and dullness. A person will rush back and forth. Despair will seize you, helplessness. Need to do something! Such a thought will torment me. This is false - Nothing needs to be done. On the contrary, go deep into inaction, freeze, and calm down if you can.

Nobody can help you in the Sidpa Bardo. All your feelings of happiness and suffering are only reflections of the bad and the good in you. Be especially careful about your feelings, hold them like horses tearing at the reins. Seeing relatives that are dragging your books, or a wife who is already intimate with another - do not try to be angry! Immediately you will fall into the underworld, from where the path is very long, there is a lot of suffering.

A person in this terrible loneliness will yearn. He will rush to look for a body to return to life again. Alas for him! While he was unconscious, while he fell from one circle of visions to another in Chonyid Bardo - his body decayed or was burned.

The resuscitation scenario is no longer possible - the body has already decomposed (we correlate with the duration of the decomposition of the flesh and the duration of the post-mortem scenario). But the desire to have a body leads to the reincarnation scenario - 1 or 2. You should be careful.

A terrible impotence will overwhelm you. Do not rush! You can throw yourself into the first womb that comes across, that will give birth you only for suffering. Be patient and careful! You may not be reincarnated at all! You can incarnate for the joys of life! There is nowhere and there is no need to rush: you are dead!

Court. Your suffering is generated by the evil in your own. If there were not evil, you would enjoy yourself here and forget about your temporary home on Earth.

If you are unable to distract yourself from your suffering, then a Good Spirit will come and begin to pile the white pebbles of your good deeds. The Evil Spirit will come with him and begin to count your bad deeds with black pebbles. Seeing these heaps, you can be frightened and begin to lie, shout that this is unfair and there were no such bad deeds, but there were more good ones. The Prince of Death will immediately appear in front of you with the Mirror of Fate (Karma), which reflects good and bad exactly.

Look into this mirror and see the truth. After that, the Prince of Death will put a noose around your neck and drag you along. Your head will be cut off, your heart will be torn out, your intestines and intestines will be scooped out and they will devour your brains, drink your blood, chew your flesh and gnaw your bones. Crazy and unbelievable pain will overwhelm you. Alas! You won't die. Your mystic body will reassemble and you will be dragged again with a noose around your neck, again torn and gnawed to pieces. This will be repeated over and over again, and there will be no end until you are aware of what is happening. Until you give an account that it is you who judges yourself, all the pictures emerge from the muddy waters of your thoughts! Neither the Prince of Death nor the monsters that tear you apart have flesh, just like you. All these terrible scenes are just haze, visions. However, here, in this world of Bardo, the vision and you are comparable in density, and therefore the pain is real!

Here, what you have created is capable of tearing you to pieces according to your terrible feelings. Don't lie when pebbles start counting. Fear not the Prince of Death! Ask him for help and protection. Recognize the whole terrible picture and do not single yourself out of it, then you will be saved! Pray to the Lord or focus your thoughts on the Great Sign of Unity. In an instant, everything will change, and you will find yourself in a safe place.

This is the last opportunity for liberation. If you fail to read this living cryptogram right now, the consciousness and memory of the past will fade. Conquer fear, be sincere at least for a moment! If you miss this opportunity, the unconsciousness of the next incarnation awaits you.

In this description of the court, we note three features: dismemberment of the body, reminiscent of shamanic initiations[4]; interpretation of judgment as occurring in itself; this is the last opportunity to avoid the reincarnation scenario. For those who did not manage to avoid reincarnation, a description of the possible worlds of birth is given:

Six worlds (lokas). *One can be born again in the six worlds. You will see their lights. The world in which you are destined to be will shine brighter than others. Smooth white light radiates the world of deities seeking pleasure; green steady light emits the world of Asuras; the world of rational human beings shines with a yellowish light; smooth blue light the world of wild animals emits, where the law of the jungle rule; the world of wandering, unfortunate spirits glow reddish and gray and black light shines from the all-cleansing Hell.*

Now that you have seen which of the six worlds shines brighter, you must realize that this is the world of your future birth and future life. Your past life will fade and melt in memory. The future will become clearer and clearer. What terrible sadness will seize you, what a despondency! Do not rush into this world of your future life that shines brighter than others. Especially if it's Hell or the Animal World. Don't despair. Try to see this light as the Light emanating from the Savior. Here are the most accurate means, the most important knowledge: gather your mind in this light, which is brighter than others, and imagine that this is the eternal Light. Let after that, the consciousness of what is around will fade away and you will go completely inside yourself. Remember, wherever there is a place, there is consciousness everywhere. Wherever consciousness arises, Dharma-Kaya immediately reveals its presence. If you manage to do this, you will slip away from birth in a bad world.

Thus:

- The Tibetan Book of the Dead emphasizes the prevention of a reincarnation scenario.
- It is opposed to a generalized post-mortem scenario, either in the form of a final convolution immediately, or starting from any stage 1-3.
- The court is understood not as a court of the gods, but as an opportunity to either complete the convolution or proceed to reincarnation, and the court

is conducted by the deceased himself with himself (reason with its binary oppositions and the will to reincarnate or complete the convolution), and this speaks of that here the deceased is at the end of transition stage 3.
- Stabilization scenario is not described; its existence can only be inferred from the mention of beings who reside permanently in the Bardo.

Let us summarize the scenarios of different epochs and cultures in Table 1, convenient for various comparisons.

Table 1. Comparison of the main post-mortem scenario with the Tibetan Book of the Dead and Commentary; + means the stage description presence in the source

Sources	Stages of the main post-mortem scenario					
	0 - initial	1 - transition	2 - choice	3 - transition	4 - distinctness	5 - merge
Tibetan Book of the Dead and Commentary	+	+	+	+	+	+

AND WHAT IS IN THE END?

So, as it was assumed in Chapter 11, the events of the post-mortem convolution of the levels and channels of the subjective space fit into several weeks, as it happens during the decomposition of the physical body. It would be possible to give a chronological reference to the stages and transitions, however, sources stipulate that the duration of events is conditional and depends on individual characteristics, so such meticulousness is inappropriate.

But an interesting question: what's next?

From the assumption of the possibility of different properties of the place where the wave goes into antiphase, in Chapter 2 three options for the development of further events were obtained: "reflection", "absorption" and "transition into antiphase". They are shown in Figure 13, chapter 2.

Let's try to complete these invariant schemes with calibrations, based on the sources used.

The reflection corresponds to the reincarnation scenario 2. The Egyptian and Tibetan Books of the Dead mention it, in Christianity such a scenario is denied. If it takes place, then it is most understandable: with the formation of a new soliton-wave pair, a new living organism is born, and the life events, described in Chapter 1, follow.

The absorption corresponds to the endpoint of fusion stage 5. Both the Tibetan and Egyptian Books of the Dead and Orthodox Christian tradition describe this moment as the approach of light - *And this is the message which we have heard from him and announce unto you, that God is light, and in him is no darkness at all* (1 John 1:5). This is followed by a difference: Buddhists and many traditional Indian schools see the ultimate goal in merging with the light and stopping samsara, the rest deny this possibility. Christians need to achieve a likeness to the light by the special grace of God. The Apostle Paul, realizing the insufficiency of words, speaks simply: *Things which eye saw not, and ear heard not, and which entered not into the heart of man, whatsoever things God prepared for them that love him* (1 Corinthians 2:9). Rev. Simeon, based on personal experience, describes this state as follows (Rev. Simeon the New Theologian, 1993, 921-922):

"The mind, being simple, or rather, naked of all thought and completely clothed in the simple Divine light...remains in the depths of the Divine light, and is not allowed to look outside at all. This is what it means: "God is light, and the highest light," which "for those who have attained it is rest from all contemplation." ... The ever-moving mind then becomes motionless and devoid of thoughts when it is completely covered with Divine darkness and light... Everything that is there is incomprehensible, inexplicable, and incomprehensible; the mind finds itself in these [realities] when it passes all that is visible and intelligible, and moves motionlessly and revolves in them, living in life above life, being light in light, but not light in itself. For then he sees not himself, but the One Who is higher than him, and, being changed by thought from the glory there, he is completely unaware of himself".

The assimilation of the Divine light is the ultimate intuitive knowledge. Neither scientific methods, nor theology, nor mystical experience can say more.

The transition to the antiphase corresponds to the reincarnation scenario 1. The potential of the wave creates an opportunity for the formation of an anti-soliton-anti-wave pair, possibly in the Anti-Universe. Chapter 2 noted that this is just a speculative assumption. Is it worth it, in the light of these sources, to understand such an Anti-universe and anti-life existence in hellish worlds? This is another assumption, the discussion of which cannot be augmented. St. John Chrysostom, answering the question of one of the parishioners, said about it this way: Why do you need to know where Hell is? You know that it exists and try to avoid it.

Thus, for scenarios of absorption and transition into antiphase, the possibilities of scientific methods are exhausted, and further discussion loses its meaning.

FUTURE RESEARCH DIRECTIONS

In Chapter 7, the dry mathematical exposition of post-mortem scenarios will be supplemented by picturesque details from low-ranking sources.

CONCLUSION

To verify the post-mortem scenario, in the absence of laboratory measurements, the only method available was used - a comparison with the universal human experience expressed in artifacts, symbols, rituals, and texts.

The test based on the Indo-Tibetan region texts leads to the same conclusion as the test based on the old artifacts and texts of the Near East. Namely: posthumous existence can be understood within the framework of the theory of complex systems self-organization as a continued evolution of open soliton-wave system components after its breakdown.

For the followers of the Tibetan branch of Buddhism, the goal is to merge with the Primordial Light and the complete cessation of the existence of the individual, and here the preservation of the body is a hindrance. But in other branches of Buddhism, incorruptible remains are a sign of holiness, and some monks subject themselves to an exhausting diet committing suicide, to achieve the incorruptibility of their bodies.

This suggests that some correlation between the state of the remains and posthumous consciousness still exists and the assumption was correct.

At the same time, it should be remembered that the information contained in the sources, is the result of meditative experience and intuitive knowledge, near-death experiences, spells of the dead, the appearances of the dead and the revived, testimonies of the living who have been in the world of the dead, the revelation of angels and gods, as well as their interpretations. These are arguments for doubting or denying the credibility of the script. The skeptic position is justified by the lack of experiments and measurements.

However, even a skeptic who believes that post-mortem experience is purely psychological (Yung, 2002) may agree with the usefulness of the presented apparatus for modeling psychological phenomena associated with ideas about post-mortem existence. As well as the usefulness of invariants, calibrations, and ternary connectives for analyzing and comparing artifacts and texts from various cultures and eras.

REFERENCES

Jung, C.-G. (2002). Psychological Commentary on The Tibetan Book of the Dead. In *Jung on Death and Immortality*. Princeton University Press., DOI: 10.1515/9780691215990-004

Kovalyov, Y., Mkhitaryan, N., & Nitsyn, A. (2020). *Self-organization of the human mind and the transition from paleolithic to behavioral modernity*. IGI Global. DOI: 10.4018/978-1-7998-1706-2

Sethumadhavan, T. N. (2011). Aitareya Upanishad Transliterated Sanskrit. Brief Explanation. (n.d.). *Transliterated Sanskrit text free translation*. Esamskriti.com. Retrieved August 15, 2024, from https://esamskriti.com/essays/Aitareya-Upanishad.pdf

Swami Nikhilananda. (Transl.). (1986-1994). Upanishads. (4 vol.). Ramakrishna Vivekanada Center

Thurman, R. (2011). *The Tibetan Book of the Dead: Liberation Through Understanding in the Between*. Bantam.

ADDITIONAL READING

Bardo Todol. [Bardo Thodol. Tibetan Book of the Dead]. Royallib.com. Retrieved August 15, 2024, from https://royallib.com/read/todol_bardo/tibetskaya_kniga_mertvih.html#0

Ksenofontov, G. V. (1992). Shamanizm. [Shamanism. Selected works.]. North-South. (Original work published 1928-1929).

Tibetskaya kniga mertvyx. (1992). [Tibetan Book of the Dead]. Chernyshev Publishing House.

ENDNOTES

[1] The reader can compare this idea with the Greek understanding of microcosm and macrocosm, and even with the self-sacrifice of Jesus Christ.

[2] Here time is measured in Earth days; posthumous sensations of time, as well as space, do not correspond to earthly ones.

³ The abilities (siddhis) mentioned in the Yoga Sutra of Patanjali are described. The ability to be in several places at the same time is inherent in advanced yogis; The Shrimad-Bhagavatam states that Krishna could exist in more than 10,000 persons.

⁴ Before the initiation ceremony, the patron spirits cut the future shaman into pieces and treat them, as well as his blood to the "masters" of diseases. The spirits cut off the head and put it on a shelf in the yurt or stick it on a stake. She retains the ability to see and hear. Having butchered the body, the spirits counted the bones of the initiate. If they were not enough, one of his close relatives had to die, giving to ransom the missing bone. After verification, the skeleton was assembled again. (Ksenofontov, 1992/1928-1929).

Chapter 7
Details From Low-Ranked Sources

ABSTRACT

Myths, legends, and fairy tales of different peoples give a variety of versions of the relationship between humans and the world of the dead. These descriptions are not scientific research, but, as can be seen from the text, even in this case there are many common features Myths, legends, and tales of different nations give the most diverse versions of the relationship between man and the world of the dead. Such texts do not have a high rating. Nevertheless, they provide interesting and colorful details about the entrances to the underworld, characterize its guardians, "lay" underground paths, and describe the "topography", as well as the rulers and inhabitants of the underworld, usually hostile to the heroes of our world. The course of the struggle for the exit from the dungeon and the methods of victory have also been preserved in the popular consciousness. The authors considered it necessary to give a summary of such descriptions.

BACKGROUND

A low rating does not mean the source's uselessness or the impossibility of scientific research.

Fairy tales and myths that lay the foundations of human behavior from early childhood. In ancient times, myths, legends, and fairy tales served as sources of poetic inspiration - (Hesiod, 2020.; Homer, n / d; Virgil, n / d., etc.), in modern times they have become the object of scientific research (E. B. Tylor, J. G. Frazer, etc.), and in our time, on their basis, quite well-thought-out theories are sometimes created (Propp, 1986; Eliade, 1981-1988; Golan, 1993, etc.).

DOI: 10.4018/979-8-3693-9364-2.ch007

Myths, legends, and fairy tales are contradictory and incompatible, which makes it almost impossible to build ternary connectives on their basis. There are other reasons why they cannot be used to verify post-mortem scenarios (Chapter 12).

Nevertheless, they contain interesting material, which can give some vitality to post-mortem scenarios.

Specifically, we will be interested in:

- the entry points to the underworld (Danylenko, 1986; Kovalyov et al., 2016; Mykhailov, 2005; Pavlova, n / d);
- the entrance guards (Bilibin, 1911);
- the underworld paths (Theudbald, 2006; Popol Vuh, 1993);
- the underworld "topography" (Parada, C. & Förlag, 1997);
- the lords and inhabitants of the underworld (Hamilton, 2009);
- a struggle with the lords of the underworld (Popol Vuh, 1993; Christenson, 2007).

Fairy tales and legends largely duplicate the material already considered, but if any reader decides to repeat the feat of Gilgamesh – he should have all the necessary information!

THE UNDERWORLD ENTRANCE

There are many variants here. Usually, this is a descent into a cave, a lake, or a well: in different parts of the Earth, such places are traditionally considered as entrances (Figure 1). Many of them have now become tourist objects.

Figure 1. Two examples of entrances to the underworld: 1 - the cave of the Sibyl (Naples, Italy), according to Virgil, the entrance to Tartarus; 2 - Aktun-Tunichil-Muknal cave (Mount Tapir, Belize), entrance to Xibalba

(Pavlova, n / d)

One of these places - the Stone Grave mentioned in Chapter 13 - was repeatedly visited by Y. Kovalyov, who once even wrote a scientific article on the psychological impact of this sacral place (Kovalyov et al., 2016). It is well-studied as a geological and archaeological object (Danylenko, 1986; Mikhailov, 2005). Several photographs in Figure 2 recreate the atmosphere of this place.

Figure 2. Stone Grave - the entrance to the world of the dead. It was used as a sanctuary for the cult of the dead from the time of the Upper Paleolithic to at least the period of Hun domination in the Northern Black Sea region: 1 - General view; 2 - Expressive details

(Kovalyov Y)

The entrance from Slavic fairy tales is Hut-on-chicken-legs. This hut, likely, demonstrates the influence of the Finno-Ugric peoples, who used it as a barn, but also had the custom of placing dead in it (Figure 3).

Figure 3. Hut-n-chicken-legs: 1 –Hut of Death; 2 - Sami barn

Sources: 1(Roerich, 2011/1909); 2(Sami barn, 2007)

THE UNDERWORLD GUARDS

The guards perform two functions - not letting the dead out, and not letting the living in. Perhaps that is why they usually have several heads. Examples of such guards: are a dog, a wolf, a snake, a scorpion-man, Baba Yaga-a bone leg, Kerber (a three-headed dog with tails in the form of snakes), a three-headed Zmey Gorynych (Dragon), and so on. They can reside in two worlds - at the gate or on the bridge.

For example, the Zmey Gorynych was waiting for the heroes on the Kalinov Bridge (from Russian - red-hot) across the Smorodina River (from Russian - currant - stink), which can be seen as a lava flow. There is a terrestrial image of this area - the Kyzylkol River and the Kalinov Bridge in the Djily-Su tract in the Elbrus region (Figure 4).

Figure 4. Kalinov bridge (a trail passes through it) on the Kyzylkol (Fiery, Smorodina) river. Elbrus region, Russia (Mysterious Kalinov Bridge - Dzhily-Su, 2021)

However, the authors, who are Kyiv citizens, truly know that the Zmey Gorynych lives in Kyiv, in the Zmievs caves on the descent to Kurenevka (Figure 5).

Figure 5. Zmievs caves. Kyiv

(Google Maps, 2024)

Let's note one more sign of belonging to two worlds - polymorphism. For example, the scorpion-man and Baba Yaga (Figure 6) are polymorphic.

Figure 6. "Here with a cheerful soul, he said goodbye to Yaga"

(Bilibin, 1911)

However, the protection is not absolute: Gilgamesh agreed with the scorpion-man, Orpheus charmed Kerber with beautiful singing, etc. But it is more difficult to agree with Baba Yaga: at first, the heroes say: Hut-hut, turn your front to me, and back to the forest; then they enter, do not let themselves be eaten or fried, sometimes they fry Baba Yaga herself, sometimes they carry out an order or agree using "personal charisma". Then Baba Yaga can help.

THE UNDERWORLD PATHS

After entering the underworld through a cave, a gate, or a hut-on-chicken-legs it is necessary to cross the seas or rivers (Acheron, Styx, Lethe, Cocytus, Phlegeton, fiery river, among the Greeks), sometimes from pus or blood, go through dense forests, wastelands or meadows, or go along a magical path or road. The Acheron River has its earthly prototype - the river of the same name in Epirus, Greece (Figure 7).

Figure 7. Acheron river

(Theudbald, 2006)

These paths lead to different parts of the underworld. Often there are multiple paths to take. To quote the Popol Vuh (Christenson, 2007):

Then went One Hunahpu and Seven Hunahpu, guided by the messengers as they descended along the path to Xibalba. They went down steep steps until they came out again upon the banks of turbulent river canyons. Trembling Canyon and Murmuring Canyon were the names of the canyons that they passed through.

They also passed through turbulent rivers. They passed through Scorpion River, where there were innumerable scorpions. But they were not stung.

Then they arrived at Blood River. They were able to pass through it because they did not drink from it.

Then they arrived at Pus River, which was nothing but a river of pus. Neither were they defeated here but simply passed through it as well.

At length they arrived at a crossroads, and it was here at the four crossing roads that they were defeated. One was Red Road and another was Black Road; White Road was one while another was Yellow Road. Thus, there were four roads. Now this, the black road said:

"Me! Take me, for I am the lord's road." Thus spoke the road.

But it was there that they were defeated. They started then on the road to Xibalba. At last, they arrived at the council place of the lords of Xibalba, and there again they were defeated.

Accordingly, the question arises: when else can you return to the world of the living? Usually, the point of no return is the crossing of the river. Indian myths add one more condition: you cannot eat the food of the dead (so as not to create a body for yourself in the world of the dead); there are hints of this in the Epic of Gilgamesh and Russian fairy tales.

THE UNDERWORLD TOPOGRAPHY

The underworld is not a single area, it is divided into several parts, in the ancient world: Tartarus, Erebus, the Champs Elysees, the palace of Pluto (Aid, Orc), around which vast deserts stretch, dull and cold, and meadows of asphodels, amazing pale, ghost-like flowers (Hamilton, 2009). The modern author presents their location in this way (Figure 8).

Figure 8. Map of the Tartar with the descents of Odysseus and Aeneas

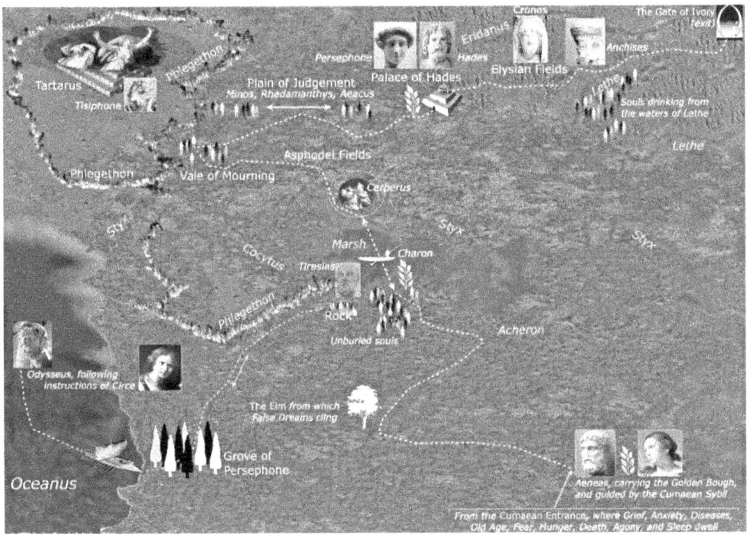

(Parada & Förlag, 1997)

And if the underworld paths, in the generalized post-mortem scenario optics, are interpreted as transition stages, then the underworld places are a different realization of the stabilization scenario. In antiquity, the residence place depended on the judges' decision. The judges' names are Rhadamanth, Minos, and Aeacus.

THE UNDERWORLD LORDS AND INHABITANTS

The lords and inhabitants are very diverse. For example, in antiquity, they called: Hades, or Pluto, and his wife Persephone, Erinius or Fury - Tisiphon, Megara and Alecto, Morpheus (Sleep) and his brother Thanatos (Death), as well as the mentioned Kerber, Charon, Rhadamanthus, Minos, Aeacus, not counting the myriad of souls of the departed; also in Xibalba, there were 12 lords - Hun-Kame, Vukub-Kame, Shikiripat, Kuchumakik, Ah-Alpukh, Ah-Alkana, Chamiabak, Chamiakhol, Kikshik, Patan, Kikre, Kikrishkak, not counting other magical creatures. Here the vowels of names are given according to (Popol Vuh, 1993); in (Christenson, 2007) the names are translated.

What do the inhabitants of the world of the dead look like? It is already clear from the given list of names that they look different. The Greek tradition depicts them as people, but the Australian aborigines drew them as shown in Figure 9.

Figure 9. Images of spirits (Aboriginal Rock Art on the Barnett River, Mount Elizabeth Station.jpg)

(Wikipedia, n.d.)

The dead unconditionally submitted to their authority, but there were conflicts with the living who penetrated the world of the dead, and the underground lords did not always win.

FIGHT AGAINST THE LORDS OF THE UNDERWORLD

This is how the Popol Vuh (Christenson A.J, 2007) talks about it. First, the lords of Xibalba defeated the descendants of the gods One Hunahpu and Seven Hunahpu. Then the twins Hunahpu and Xbalanque, miraculously born from the head of One Hunahpu, came to take revenge on them. They crossed all the rivers and came to the four roads.

When they came to the four crossroads, they already knew the roads of Xibalba – the Black Road, the White Road, the Red Road, and the Blue/Green Road.

Then they sent an insect named Mosquito. They sent him on ahead to obtain for them what he could hear:

"You shall bite each one of them in turn. Bite the first one seated there and then keep on biting them until you have finished biting all of them. It will be truly yours then to suck the blood of people on the road," the mosquito was told.

"Very well then," said the mosquito.

So, then he went along the Black Road until he alighted behind the effigies of carved wood. The first ones were all dressed up. He bit the first one, but there was no response. Then he bit the second one seated there, but he did not speak either.

Next, he bit the third one seated there, who was One Death – "Ouch!" said each one when he was bitten.

"What?" was their reply: "Ow!" said One Death. "What, One Death? What is it?"
"I am being bitten!"

So, the twins learned the real names of all the lords of Xibalba.
Thus, they came to where the Xibalbans were.
"Hail these lords who are seated there," said a tempter.
"These are not lords. These are merely effigies carved of wood," they said. Then they hailed each one of them:

"Morning, One Death. Morning, Seven Death. Morning, Flying Scab. Morning, Gathered Blood. Morning, Pus Demon. Morning, Jaundice Demon. Morning, Bone Staff. Morning, Skull Staff. Morning, Wing. Morning, Packstrap. Morning, Bloody Teeth. Morning, Bloody Claws," they said when they arrived there.

All of them had their faces revealed, for all of their names were named. Not one of their names was missed. When they were called upon, they gave the names of each one without leaving any of them out.

"Sit down here," they were told, for it was desired that they sit on top of the bench. But they didn't want to:

"This isn't a bench for us. It is merely a heated stone," said Hunahpu and Xbalanque. Thus they were not defeated.

"Very well then, just go into that house," they were told.

So then they entered into the House of Darkness. But they were not defeated there.

"What becomes of them? Where did they come from? Who begat them? Who gave them birth? Truly our hearts are troubled, for it is not good what they are doing to us. Their appearance as well as their nature are unique," they said one to another.

Then they summoned all of the lords:

"Let us play ball, boys," they were told.

But first they were questioned by One Death and Seven Death:

"Where did you really come from? Would you tell us, boys?" they were asked by the Xibalbans.

"We must have come from somewhere, but we don't know." Only this they said. They told them nothing.

"Very well then. We will just go and play ball, 378 boys," the Xibalbans said to them.

"Fine," they replied.

"Here is our rubber ball that we will use," said the Xibalbans.

"No, we will use ours," said the boys.

In the end, they got what they wanted:

"Very well then," they replied.

Thus they took out their rubber ball, and it was thrown down.

Then the twins successfully pass all other tests. Finally:

There the Xibalbans wanted to force them into playing with them:

"Let us jump over this our sweet drink. Four times each of us will go across it, boys," they were told by One Death.

"You cannot trick us with this. Do we not already know the means of our death, O lords? You shall surely see it," they said.

Then they turned to face one another, spread out their arms and together they went into the pit oven. Thus both of them died there.

Then all the Xibalbans rejoiced at this. They contentedly shouted and whistled:

"We have defeated them. None too soon have they given themselves up," they said.

Then they summoned Descended and Ascended, with whom word had been left by the boys. And the Xibalbans divined of them what was to be done with their bones. Thus according to their word, the bones were ground up and strewn along the course of the river. But they did not go far away; they just straightaway sank there beneath the water. And when they appeared again, it was as chosen boys, for thus they had become.

On the fifth day they appeared again. People saw them in the river, for the two of them appeared like people-fish. Now when their faces were seen by the Xibalbans, they made a search for them in the rivers.

And on the very next day, they appeared again as two poor orphans. They wore rags in front and rags on their backs. Rags were thus all they had to cover themselves. But they did not act according to their appearance when they were seen by the Xibalbans. For they did the Dance of the Whippoorwill and the Dance of the Weasel. They danced the Armadillo and the Centipede. They danced the Injury, for many marvels they did then. They set fire to a house as if it were truly burning, then immediately recreated it again as the Xibalbans watched with admiration.

Then again they sacrificed themselves. One of them would die, surely throwing himself down in death. Then having been killed, he would immediately be revived. And the Xibalbans simply watched them while they did it. Now all of this was merely the groundwork for the defeat of the Xibalbans at their hands.

At length the news of their dances came to the ears of the lords One Death and Seven Death. And when they had heard of it, they said: «Who are these two poor orphans? Is it truly delightful? Is it true that their dancing and all that they do is beautiful?"

At length they arrived before the lords. They pretended to be humble, prostrating themselves when they came. They humbled themselves, stooping over and bowing. They hid themselves with rags, giving the appearance that they were truly just poor orphans when they arrived. Then they were asked where their home mountain was and who their people were. They were also asked about their mother and their father:

"Where do you come from?" they were asked.

"We do not know, O lord. Neither do we know the faces of our mother or our father. We were still small when they died," they just said. They didn't tell them anything.

"Do not be afraid or timid. Dance!" [Xibalbans] were told.

Thus they began their songs and their dances, and all the Xibalbans came until the place was overflowing with spectators

And the lords marveled at it:

"Now sacrifice yourselves. We would see this. Truly it is the desire of our hearts that you dance," *said again the lords.*

"Very well then, O lord," *they replied.*

So then they sacrificed themselves. Hunahpu was sacrificed by Xbalanque. Each of his legs and arms was severed. His head was cut off and placed far away. His heart was dug out and placed on a leaf. Now all these lords of Xibalba were drunk at the sight, as Xbalanque went on dancing.

"Arise!" *he said, and immediately he was brought back to life again. Now the lords rejoiced greatly. One Death and Seven Death rejoiced as if they were the ones doing it. They were so involved that it was as if they themselves were dancing.*

For it was the desire of the lords to abandon their hearts to the dances of Hunahpu and Xbalanque. Then came the words of One Death and Seven Death:

"Do it to us! Sacrifice us!" they said.

"Very well then. Surely you will be revived. Are you not death? For we are here to gladden you, O lords, along with your vassals and your servants," they said therefore to the lords.

The first to be sacrificed was the very head of all the lords, One Death by name, the lord of Xibalba. He was dead then, this One Death. Next they grabbed Seven Death. But they didn't revive them. Thus the Xibalbans took to their heels when they saw that the lords had died. Their hearts were now taken from their chests. Both of them had been torn open as punishment for what they had done. Straightaway the one lord was executed and not revived. The other lord had then begged humbly, weeping before the dancers. He would not accept it, for he had become disoriented:

"Take pity on me," he said in his regret.

Then all of their vassals and servants fled into the great canyon. They packed themselves into the great ravine until they were piled up one on top of the other. Then innumerable ants swarmed into the canyon, as if they had been driven there. And when the ants came, the Xibalbans all bowed themselves down, giving themselves up. They approached begging humbly and weeping. For the lords of Xibalba were defeated. It was just a miracle, for the boys had transformed themselves before them.

And then they declared their names. They revealed their names before all Xibalba.

So, the rules of this magical game, where the stake is final death, are as follows: make the opponent take the illusion for reality, fool him, and sacrifice him, already defenseless.

And the victory algorithm is as follows: do not give the enemy any information about yourself, especially about your name, find out the name of the enemy, do not succumb to his deceptions, distinguishing between truth and illusions; impose their rules and their will on him; to voluntarily sacrifice oneself, having received new opportunities in return; fool him, and, having won, sacrifice him without any mercy.

Figure 10. The Mother of the Hun-Akpu and Xbalanque before the lords of Xibalba

(Popol-Vuh, 1993)

FUTURE RESEARCH DIRECTIONS

In addition to the natural science data on the problem of death and the modeling results, we present in Chapter 8 a philosophical view of the problem of individual finitude.

CONCLUSION

Myths, legends, and fairy tales of different peoples give very different versions of the relationship of a human with the world of the dead. But, as can be seen from the text, even in this case there are many common characteristics: the world of the dead is located underground, it has an entrance preserved by the guards, the transition is associated with the crossing of rivers; the world of the dead is a world of deception, illusions, and magic, and its masters can be defeated.

Let us note again, that the descriptions given above are not scientific.

REFERENCES

Christenson, A. J. (2007). *Popol Vuh: The Sacred Book of the Maya*. University of Oklahoma Press.

Google Maps. Kyiv. https://www.google.com.ua/maps/search/zmieva+pechera+Kiev/@50.4574117,30.5183514,12.79z?hl=ru&entry=ttu

Hesiod. Theogony. Nagy, G.& Banks, J., (Trans.). (2020, November 2). The Center for Hellenic Studies. https://chs.harvard.edu/primary-source/hesiod-theogony-sb/

Homer. *The Iliad*. Butler, S. (Trans.). (N.d.). Mit.edu. Retrieved August 15, 2024, from http://classics.mit.edu/Homer/iliad.html

Parada, C., & Förlag, M. (1997). Map of the Underworld. Showing the descents of Odysseus and Aeneas. [Illustration]. *Map of the underworld - Greek Mythology Link*. (n.d.). Maicar.com. Retrieved August 15, 2024, from https://www.maicar.com/GML/Underworldmap.html

Theudbald. (2006). River Acheron. [Illustration]. Date: 2006. Author: Theudbald. (N.d). Wikimedia.org. Retrieved August 15, 2024, from https://commons.wikimedia.org/wiki/File:Acheron.JPG?uselang=en

Virgil. *The Aeneid*. Dryden, J. (Trans.). (N.d.). Mit.edu. Retrieved August 15, 2024, from http://classics.mit.edu/Virgil/aeneid.html

Wikipedia contributors. (n.d.). *File: Aboriginal rock art on the Barnett River, Mount Elizabeth Station.jpg*. [Illustration]. Author: Graeme Churchard. Wikipedia, The Free Encyclopedia. Retrieved September, 2, 2024. https://en.wikipedia.org/wiki/File:Aboriginal_rock_art_on_the_Barnett_River,_Mount_Elizabeth_Station.jpg

ADDITIONAL READING:

Bilibin, I. Ya. (1911). *Vot s veseloj dushojy on poproshhalsja s Jaghoy*. [Here with a cheerful soul, he said goodbye to Yago. Illustration for "The Tale of the Three Royal Divas and Ivashka, the Priest's Son"]. *File: Ivan Bilibin 036.jpeg*. (n.d.). Wikimedia.org. Retrieved August 15, 2024, from https://commons.wikimedia.org/wiki/File:Ivan_Bilibin_036.jpeg?uselang=ru

Danylenko, V. M. (1986). *Kamyana Mohyla* [Stone tomb]. Naukova Dumka.

Eliade, M. A (1981-1988). *History of Religious Ideas*. (3 volumes). University of Chicago Press.

Golan, A. (1993). Myf i simvol. [Myth and symbol]. *Russian Literature*.

Hamilton, E. (2009). *Myfy y leghendы Ghrecyy y Ryma* [Myths and legends of Greece and Rome]. Tsentrpoligraf.

Kovalyov, Y. N., Nitsin O. Y. & Shevel L.V. (2016). Modeljuvannja i ocinjuvannja vzajemodiji ljudyny iz sakraljnym seredovyshhem (na prykladi NIAZ «Kam'jana Moghyla»). [Modeling and evaluating the interplay of people with the sacred environment (on the butt of NIAZ "Kamyana Mogila")]. *Modern problems of modeling* 5,.57-65

Kuhn, N. (1922). *Leghendy y myfy Drevnej Ghrecyy*. [Legends and myths of Ancient Greece]. Public Domain. https://bookscloud.ru/books/80264

Mykhailov, B. D. (2005). *Petroghlify Kam'janoji Moghyly: Semantyka. Khronologhija. Interpretacija* [Petroglyphs of Kamyana Mohyla: Semantics. Chronology. Interpretation]. MAUP.

Pavlova, M. (2017, March 15). *10 skrytykh vorot v podzemnyj myr*. [10 hidden gates to the underworld]. [Illustration]. Flytothesky.Ru. https://flytothesky.ru/10-skrytyx-vorot-v-podzemnyj-mir/

Popol-Vuh. (1993). Nauka. [Popol-Vuh]

Propp. V. Ya. (1986). *Ystorycheskye korny volshebnyh skazok*. [The historical roots of fairy tales]. Publishing House of Leningrad State University

Roerich, N. K. (1909). Izba smerti. [Hut of Death]. [Illustration]. (N.d.-f). Wikimedia.org. Retrieved August 15, 2024, from https://commons.wikimedia.org/wiki/File:Nicholas_Roerich_-_hut_of_death.jpg?uselang=ru

Saamskyj saraj. (2007). [Sami barn]. [Illustration]. Model in the Skansen Park (Stockholm). Sami Storehouse on stilts, displayed at Skansen in Stockholm. Date: 10 July 2007. Author: m.prinke. *File: Sami storehouse.Jpg*. (n.d.). Wikimedia.org. Retrieved August 15, 2024, from https://commons.wikimedia.org/wiki/File:Sami_Storehouse.jpg

Taynstvennyj Kalynov most - Dzhyly-Su. (2021). [Mysterious Kalinov Bridge - Dzhily-Su]. [Illustration]. (N.d.). Trip8.Ru. Retrieved August 15, 2024, from https://trip8.ru/zagadochnyy-kalinov-most-jily-su/

Chapter 8
The Transpersonal Experience in Religion and Culture as a Response to the Challenges of Death

ABSTRACT

The genesis of ideas about individual finiteness in Western philosophy is shown: death is an initiatory event, the meaning of which depends both on the state of human consciousness in each of the historical eras, and on the characteristics of individual consciousness. In the postmodern period, individualism, material values primacy, ignoring transcendental things, tabooing the problem of death, and the loss of meaning prevailed in everyday consciousness. The way out of the crisis is the individual attitude to the sacred secret of being, capable of filling a person's existence with meaning, significance, and happiness. The highest manifestation of spirituality is transpersonal experience, which initiates a person's approach to his essence through symbolic death and rebirth. The reality of the absolute, revealed in experience, is comprehended not rationally discursively, but intuitively contemplatively, through existential involvement.

BACKGROUND

The origins of modern Western philosophy ideas about the life cycle problem, the emergence of life, the meaning of being, and death, arose back in Antiquity. This era was characterized by a transition from mythological to rational conscious-

DOI: 10.4018/979-8-3693-9364-2.ch008

ness (Kovalyov et al., 2020), as well as a highly creative activity. Accordingly, the comprehension of mythologemes and initiatory rites, the interpretation or denial of individual plots and images gave rise to a variety of philosophical views and schools characteristic of this era, a variety that permeated all ancient literature and culture as a whole.

The ideas generated by Antiquity and expounded by Pythagoras, Socrates, Plato, Aristotle, Euclid, Plotinus, Proclus, and many others continue to feed modern philosophy; in particular:

- Various aspects of the problem of interpretation and verification of symbolic texts are developed (Bakhtin, 1979; Korshunov, 1991; Losev, 1982; 1993; Lotman, 2000; Van Dijk, 2006; Gadamer, 1989).
- Devoted to understanding transpersonal experience (Wilber, 2008; Maslow, 1968; Eliade, 2002; Torchinov, 1997; Heidegger, 2006/1927; Wittgenstein, 1995), in particular, the importance of inspiration in creativity and overcoming the limitations of the individual (Heidegger, 2012).
- The interpretation of being as a series of initiations, including the event of death, the state of profanation, the process of desacralization, and the role of moral factors are explored (Frank, 1990; Morozov, 2009; 2018; Tillich, 1995; Otto, 1954/1929; 1995/1933; 2015/1956).

The Middle Ages, which replaced Antiquity, was characterized by the dominance of the Christian paradigm in the West, with some infiltration of ancient, Jewish, and Islamic content. Since such infiltration took place mainly through the mediation of various Hermetic communities, one can say more definitely about its content - the magical ideas of the Ancient World, Kabbalah, and Sufism passed through the "filter" of these communities. The study of traditional Christian ideas and infiltrated content in the aspects that interest us is devoted (Guénon, 2004; James, 1993; Minin, 2003; Florensky, 1993; Schmemann, 2006).

Modern times and modernity (especially since the 19th-20th centuries) are characterized, among other things, by the wide availability of sources on religion, mysticism, magic, and philosophy throughout the globe and the entire culture of mankind (Mercer, 1952; Wallis Budge, 1967/1895; Radhakrishnan, 2009/1923; Dasgupta, 2014/1922). However, Western philosophy, enriched with individual ideas (let us single out reincarnation and transpersonal experience), has developed its terminology and methodology for dealing with them, without losing its identity. It can even be said that only those ideas were accepted that had their roots and analogs (reincarnation - in Antiquity and Kabbalah; transpersonal experience - in Christian and Sufi mysticism, and so on).

In contemporary Western philosophy a traditional ethical and metaphysical analysis of death and fear of death is carried out (Taylor, 2014; Yourgrau, 2019; Fischer, 2019; Hagglund, 2019), because death stands at the center of profound moral questions of the meaning of life, liberty, non-existence, and life after death as well. Meanwhile, various thinkers concentrate their attention not on metaphysics or religious ethics but rather on emotions and attitudes toward death and mortality such as fear, existential terror, horror, regret, anger, resentment, disorientation, stress, and even gratitude (Cholbi, 2016, 2022; Behrendt, 2018). The problems of psychological adaptation of dying people and their families (Kamm, 2021; Kaufman, 2022), and the ways of overcoming the grief and sorrow that always surround death (Cholbi, 2023), are also raised. Many interdisciplinary works try to look at philosophical and existential aspects of death from the point of view of biology and medicine (Charlier, 2018; Manabu Fukuda, 2023). While Western European tradition of philosophical thought tends to continue a rational Epicurean atheistic argumentation toward fear of death Kokosalakis, 2020; Kaufman, 2016), non-western thinkers based on orthodox Christian tradition and 'philosophy of heart' tend to regard death as the metaphysical mystery that helps a person to transcend oneself and pass through an initiation-travel to a higher spiritual state of mind (Kovalyov et al., 2020; Morozov, 2022, 2024; Torchinov, 2015, Schmemann, 2006).

GENERAL VIEW

The individual's desire to find one`s existence meaning, to achieve the spiritual and meaningful fullness and significance of life, and to fulfill and realize oneself is primordial and universal. Nothing paralyzes the freedom of the human personality as strong as the disturbing feeling of absurdity, meaninglessness of one's existence, and devaluation of the world. Sometimes this feeling can be especially acute at the moment of illness and death. In turn, the "will to meaning", as an innate human orientation and need, in its extreme manifestations is closely related to the experience of higher, unconditional values of goodness, beauty, happiness, and bliss. The latter is accompanied by a feeling of boundless joy, love, empathy for everything that exists as a good and a gift, awareness of the infinity of the individual, the illusory nature of death, etc.

The experience of happiness-bliss is a kind of indicator of how the will to meaning was successfully embodied. And vice versa: depression, frustration, apathy, suicidal states, escapism and avoidance of reality (alcoholism and drug addiction), unconscious fear of death, nihilistic rejection of cultural traditions, indifference or ironic attitude to the great tasks of culture, its classical values and ideals, mistrust of "great narratives" (F. Lyotard) - all this indicates the defeat or weakness of the will to meaning.

The connection between the concepts of "meaning of life" - "values/virtues" - "happiness-bliss - "and fear of death" used to be self-evident for a premodern man, or "homo religious" (Eliade, 1981). Here one needs to understand the premodern background culture. However, the latter is not homogeneous. Due to different types of premodern cultures, various interpretations of the above-mentioned connection are possible. Is happiness-bliss without anxiety (and without the fear of death as one of the manifestations of anxiety) - this shining peak of human self-realization and the goal of intelligent existence - the result of righteousness and living according to virtue alone? And are ethics alone (or even formal rules of etiquette) enough for us to realize this higher purpose of life? The ancient Chinese and ancient Greeks would give an affirmative answer. And here fearlessness in the face of death is also a completely rational and learned ("trained") thing, which depends on a person's willpower.

However, another option is possible, when happiness-bliss does not depend solely on human efforts and requires an "unworldly" source, for example, what is called in Christianity or Islam a spiritual "prayer experience", which is given by the grace of the Lord "not of this world". In this case, fearlessness in the face of death turns out to be the result of other, supernatural, inhuman forces. In other words, the question arises: is the completeness of a person's spiritual development a natural state, i.e., the realization of the inherent capabilities of human nature, literally "*self*-realization"? Or is it a supernatural state (that is, where a specific transpersonal/transcendental experience is needed, in which the Self is overcome)? So, is it enough for a person to be happy and get rid of the fear of death, what nature gives him, or does he strive with all the strength of his soul for something greater, supernatural?

DISADVANTAGES OF THE "IMMANENT PARADIGM"

If use the "immanent paradigm" without appeal to transcendental dimension, then this epistemological strategy sooner or later (through a series of historical and philosophical mediations) will lead us to a materialistic understanding of nature in general and the materialistic nature of consciousness in particular, to the inevitable conclusion that happiness is an epiphenomenon of complex biochemical processes occurring in the brain. Thus, the solution of philosophical issues (meaning of life, moral goodness) within the limits of this strategy is transferred to the vulgar materialistic (positivist) plane of intersection of psychiatry, neurophysiology, dietetics, and pharmaceuticals, that is, the corresponding technique of "normalization" of brain biochemistry. Today's level of science makes it possible to develop medicines for almost all symptoms of the "existential vacuum" (depression, anxiety, apathy, aggression) which were reported by V. Frankl, E. Fromm, P. Tillich, R. May, K. Jung, and other researchers of the humanistic direction in psychology and theology. One takes the appropriate pill - and becomes "happy", feeling excitement, euphoria, cheerfulness, fearlessness, and boundless love for everyone. When the effect of narcotic drugs has ended, the essential fullness of happiness-bliss is rolled back to the ordinary boredom and absurdity of everyday life, the realm of the impersonal ("Das Man"). It is clear that such therapy is symptomatic, but does not treat the spiritual cause of the disease, because here the causal relationship between happiness and the virtues necessary for its achievement is broken (Morozov, 2024).

The achieving happiness technique within the "immanent paradigm" limits can have a more complex, social-engineering character. With the help of manipulative technologies, a stereotype is introduced into the mass consciousness, according to which happiness is equal to a certain set of "signs of happiness". This is how a "normal" conformist person who plays by the rules of the game of consumer society is formed. Happiness is reduced to things and services that demonstrate a certain status of a person in society: clothes, a car, a house, gadgets, etc. Respectively, an unhappy person is a loser, who for some reason lacks "signs of happiness." The situation is even worse for someone who does not see the point in receiving "signs of happiness" - such a person is abnormal, and deviant for the ideology of consumerism. This is regarded to be criminal or marginal behavior. E. Fromm in his "To Have or to Be?" quite carefully described the falsity, futility, and vulgarity of a hedonistic attitude to acquisition, possession, and accumulation. Therefore, we will not specifically focus on this.

BEYOND THE FACTS

Speaking about the meaning of life and death within the "immanent paradigm", we need to refer to Wittgenstein's famous theses that "the world is the totality of facts, not of things" and "facts in logical space are the world" (Wittgenstein, 1999). This thesis immediately faces the problem of explaining the symbolic forms of culture in which a person lives, as the value dimension. After all, it is impossible to empirically derive norms, quality, meaning, and value from the facts. Values are counterfactual. Human existence is an empirical "fact", but what is the meaning and value of this fact? "Fact" and "value of fact" (and ultimately - "meaning of fact") belong to different orders of being. This can be compared to the difference between two-dimensional and three-dimensional space.

Values are not and cannot be the subject of positive science. This is a fundamental non-scientific field of competence. L. Wittgenstein writes:

«6.4. All propositions are of equal value.

6.41.   The sense of the world must lie outside the world. In the world, everything is as it is and happens as it does happen. In it there is no value —and if there were, it would be of no value. If there is a value that is of value, it must lie outside all happening and being so. For all happening and being-so is accidental.

What makes it non-accidental cannot lie in the world, for otherwise this would again be accidental. It must lie outside the world.

6.42. Hence also there can be no ethical propositions. Propositions cannot express anything higher.
6.421.   *It is clear that ethics cannot be expressed»*. (Wittgenstein, 1999)

If all the facts about the world became known, the main questions of life (ethical, worldview questions) would not even be raised. If moral values are counterfactual, then science cannot say anything about them. One can only be silent about values, indicating their place in the logical structure of the world and the cognitive process: *"6.422. Ethics is transcendental. Ethics and aesthetics are one"* (Wittgenstein, 1999).

It would seem that such a statement has purely negative consequences from an ethical point of view and has no logical meaning because the key metaphysical concepts remain scientifically unproven/unexplained: "evil", "good", "humanity", "moral", etc. So, the positivist picture of the world, making strict ethical judgments impossible from a scientific point of view, one way or another leads us to a situation of moral nihilism (Morozov, Nikulchev, & Kulagin, 2021). At the same time, positivism shows that the "main questions of life" («to be or not to be», «is life worth living») remain metaphysical questions and are accessible in a special non-discursive way.

"7. Whereof one cannot speak, thereof one must be silent" (Wittgenstein, 1999).

It is clear, that the silence Wittgenstein is talking about is not a logical denial of speech, not just the absence of sound/noise or an artificial pause in a conversation when the interlocutors have nothing to say. The silence here means a specific experience of spiritual contemplation, where unspeakable (non-verbal) things are intensely experienced. How and what exactly is experienced in the state of silence is not discussed in the "Tractatus", but only hinted at. It can be assumed that the last short aphorism of the Tractate contains an implicit reference to Platonism, the Byzantine tradition of hesychasm, and Buddhist prayer states. From a methodological point of view, the limits of scientific competence must be shown (answering Kant`s question "What can I know"), and at the same time, the existential (value-meaning) sphere is outlined, where non-scientific forms of knowledge (religion, mysticism, poetry) reign. Logical positivism denies metaphysics as a science but does not deny its subject field. It is from this methodological construction, that a further path to those trans-personal forms of cognition can be outlined, which overcomes the "closed shell of immanence" and lead a person into the transcendent, counter-factual realm.

THE TOTALITY OF THE TEXT

The further "immanent paradigm" development leads us from logical positivism to the final deconstruction of metaphysics in postmodernism. A fact is a certain event that can be described in a text - written or oral. Facts make up a certain story, a narrative that "responds" or co-relate to another narrative, and enters into a dialogue with it (confirms, denies, doubts, believes, ironizes, re-evaluates, etc.). To say that "the world is a collection of facts" is essentially the same as saying that "the world is a collection of texts." The latter make up a certain cultural-historical context, which, in turn, leads a dialogue with other cultural-historical contexts, and so on ad infinitum. These giant archipelagos of texts are autonomous systems in which, according to the post-structuralism paradigm, the "signifier" is separated from the "signified" (the objectivity of things), and "the author dies" (R. Barthes). The author's monopoly position on the only correct interpretation of the text is denied, and any semantic hierarchy (truth-false, better-worse, sublime-vulgar) is eliminated. In this totality of the text ("there is nothing outside the text") specific values and norms are declared conventional, discursive, historically and culturally determined, existing, and generated in specific culture time-space.

However, to claim that values are only contextual and discursive means the destruction of ethics, ending in relativism and nihilism. Just think about it! In one text, Hitler is Prometheus, the messiah of the Aryan race, in the second one - the curse of Europe and the executioner, in the third one - a comic book character, in the fourth - the personage of a song, and so on. In one discourse, the war is a harsh and tragic reality, in another, the same war is a farce and hoax, in the third, it is the subject of an anecdote from the "black humor" series. In this "polyphonic" variety of positions, points of view, and "voices" the author's meta-position sinks: since all these texts are of equal value, there is no truth in the texts. No one can claim truth, universality, or non-involvement. If values are texts that are born in other texts, then the postmodern model of the "rhizome" (J. Deleuze) is suitable for describing the symbolic world. In this labyrinth, there are no classical metaphysical oppositions of center and periphery, spiritual "heights" and "lows", good and evil (Morozov,

2024). From a psychiatric point of view, such a rhizome indicates a schizophrenic diagnosis, a pathological mental process, where the "Ego" is split as a transcendental center and no longer can provide the unity of the individual. Instead of this center, various "personas" begin to compete with each other (C.-G. Jung).

Therefore, positivist and post-structuralist worldviews make universal, objective, and necessary moral values impossible to exist. If good, evil, human nature, the meaning of life, freedom, and eternity are metaphysical, non-empirical concepts, or contextually determined, relative, and changing "discursive arrangements", we cannot operate with any moral and legal categories claiming universality, such as "crime against humanity", "crime without a statute of limitation", "humanism", "natural rights", "conscience", "honor", "dignity", etc. If the historical "texts" each time can be rewritten and interpreted differently, there is no sense in the existence of museums commemorating the genocide, or the victims of Hiroshima and Nagasaki, funds for the protection of the rights and freedoms of citizens, and so forth. When universal values are denied, the world we live in falls apart, and all political rhetoric about democracy, human rights, and humanism "fall apart" as completely metaphysical (unscientific and unproven). It can be sharpened that if the world is texts, contexts, and discourses, then its ontological (real) status is called into question. In such optics, the world of things is a great illusion, Hindu Maya. However, the logical inconsistency of such ontological schizophrenic "splitting" is obvious. The statement "there is no universal value constants" must fall under its requirement, so it cannot be universal. When it is declared that "everything is relative and historical", then this declared thesis is relative and historical. The thesis that "there are no objective things, but only texts" - also belongs to the text, and therefore to the context, and cannot claim truth, objectivity, or necessity.

"BACK TO THINGS"

To avoid such extremes of nihilism and to find an opportunity to operate with universal and objective values, it is necessary to return to what is outside of reality and the text (which means outside of theories, concepts, intellectual superstructures, and models) - "back to things" (E. Husserl). The world is not only facts, events, subjects of thought, and texts but, above all, a "collection of things". It is clear that in the case of meanings and values, we are talking about "things" in the ideal sense of the word. In Platonism, things are "eidos", in phenomenology - "phenomena of consciousness" (noema), in the philosophy of J. Santayana - the "realm of essences", in the system of A. Whitehead - "ideal objects", in J. Derrida`s one - hypothetical "transcendental signified" etc. For Heidegger, things are a special kind of being; we can evaluate them through the "fourfold of being", formed by pairs of opposites:

heaven - earth, mortal people – immortal gods. Through things the "truth of being" speaks to us, and through the category of "being" you can go to the next category of "good" because everything that exists by its involvement in the truth of being is good. The metaphysical triangle of unity-truth-good enables all further philosophical and theological attempts to build a value hierarchy. Values and meanings are things that speak, "speak to us from above", and call for their embodiment. At the same time, things that are not facts and are outside the text require a special vision-understanding, that is, a special intuitive transpersonal experience of insight into their essence. Without this experience, we are in "value blindness" (M. Scheler), unwilling to respond to their challenge.

As for the thesis about the historical determinism of values and anti-universalist tendencies, it can be answered that the very possibility of distinguishing the orders of being – what is and what ought to be facts and values - is universal and meta-historical. That is, the possibility of distinction is not culturally determined but acts as an a priori condition for the possibility of any culture in general. The ability to endow facts with meaning and value is a kind of transcendental grammar of culture, without which facts would have no meaning. Facts are "ordered" (explained, interpreted, evaluated) only with the help of this transcendental grammar. And this grammar, in turn, would not be valid if it were not based on the objectivity of things, and hence the objectivity of their values. (Here one can refer to the philosophy of M. Scheler and N. Hartman, who used the concept of "material a priori" in ethics and advocated the principle of an objective hierarchy of values).

The sphere of the transcendental is the prerogative of metaphysics. Therefore, the problematic of the meaning and value of existence as such, of the human "will to meaning", leads us, one way or another, beyond the boundaries of "nature" and "facts", to the path to religion, which is the quintessence of metaphysical (spiritual) searches of humanity.

SPIRITUALITY

Just as for the "immanent paradigm" the state of happiness, satisfaction, and tranquility is an epiphenomenon of material (biological, physical, chemical processes in the body), so for the religious ("supernatural") paradigm, they are also interpreted as derivatives, by-products - but no longer of material processes, but of a certain spiritual work and spiritual experience, from which they receive their significance. Hegel's remark that "the secret of happiness lies in the ability to leave the circle of one's self" (Hegel, 1977) is important. We can take it as a starting point for further consideration. This means that a person locked in his egoism feels his loss and lack of self-sufficiency. And the more the ego identifies itself with the body, the more painful and intense these sensations are. If I am my body, am meant to be in the body. The death of the body is the cessation of the individual's existence. Only by opening the circle of the individual world in the act of transcendence (overgrowing individual boundaries), overcoming self-centeredness and monologue, overcoming fixation on oneself, above all on one's body, becoming a part of the whole, does a person achieve essential completeness and peace of mind. Realizing that I am not just a body that will die, but a spirit, an ideal entity free of space and time, is an important step in this process of unfolding. The "eccentricity" of human existence, placing the center of meaning not in oneself, but outside, shows its dependence on the Other, the futility of all efforts to build a monad-like autonomous "selfness". At the same time, the Other, his "face" (E. Levinas), in his inner world is infinitely different from the I. The Other is always absolutely Other. Such absolute "otherness" of our neighbor next to us inevitably leads us to the monotheistic concept of God-Personality, absolutely Other than the world created by him. Thus, we gradually approach the question of the meaningful center of gravity for a person who, overcoming the "circle of his ego" (Hegel), seeks meaning and happiness that would not depend on the fact of physical death. Here a completely different problem arises - spiritual life and spiritual death. If the biological death of the body is inevitable and necessary, then excessive "investment" in what is supposed to disappear with the disappearance of the body turns out to be useless. Therefore, in religion, the emphasis is on awakening a person from his spiritual sleep, to prevent spiritual death in eternity.

The very word existence (existentia) is related to the word ecstasy. Ecstatic existence indicates it draws life force not from itself, but from an external source, and secondly, that the constant process of self-overcoming ("coming out of oneself") has an absolute semantic horizon. In other words, if existence is transcendence, then there is also "Whereto-and-Wherefrom of Transcendences", the sacred and ineffable mystery of being.

Personality is born in the participation of the individual with the supra-individual. Only in this longing for the "peaks of being" does it grow and expand in value. And vice versa: existential refusal to serve super-individual meanings and goals, indifference to "higher" super-tasks, lack of "taste for eternity" (that's how Schleiermacher defined the essence of religious consciousness) ends with the stagnation of "personhood" as a principle. The more a modern secular person is focused on "his", purely individual, particular, the less happy he is, and the more often he feels alienation, loneliness, and existential vacuum, which he often compensates by unrestrained hedonism, drugs, unmotivated aggression, moral and legal nihilism, sectarianism, virtual reality, etc.

What is spirituality? Today, the concepts of spirit and spirituality are often used figuratively and allegorically, blurring their original meaning more and more. By "spiritual" here and further, we understand the manifestation of being as a sacred mystery of the divine in the everyday world and the symbolic form of this mystery expression in personal experience. In other words, spirituality is a person's attitude towards the sacred (Morozov, 2022). The latter, due to this attitude, reveals to a specific person in the form of an epiphany ("revelation"). Without this phenomenon-epiphany, the transcendent realm of the sacred would always remain an impenetrable mystery, an unknowable "thing-in-itself." The denial of epiphany (the possibility of experiencing the divine) leads to the extremes of Kantian agnosticism and the subsequent anti-metaphysical nihilistic line of philosophizing. Thanks to the spirit (Greek "nous"), the personality is an ontologically open structure, to which the sacred secret of existence is revealed. If we use traditional terminology, spiritual experience is a gap in a wall through which the light of divine energies enters the darkness of the material world, filling up human life with a higher meaning. In more modern terms, it can be called a "peak experience" (A. Maslow), a goal experience in which a person reaches the absolute horizon of transcendence, the "final" answer to fundamental existential questions. In this peak experience, a person opens up a vision of the things' essence - as they are in their essence, as well as the true meaning of their existence. In turn, an unspiritual man lacks the sacred and insensitive to things of a super-material order, and therefore a descent into vulgarity. Unspirituality is a "meta-pathology" of personality (A. Maslow, 1968).

Transpersonal experience exalts a person's existence, connecting it through a steps with the mystery of the sacred and "incomprehensible" (S. L. Frank). Such an experience, unlike abstract knowledge, qualitatively changes a person's life. Transpersonal experience, or the experience of being (in this case we will consider them as synonyms) is constitutive in the process of becoming a person, his humanization. Without it, a person loses his privileged ontological status, dehumanizes, and turns into a thing among other things, human material. At the same time, the "transcendence" of the individual takes place in this experience, which implies a constant spiritual movement of improvement and self-overcoming. This "upward movement" can also be called spiritual initiation, that is, symbolic death and resurrection, through the gradual moral transformation of the individual. For example, the Byzantine ascetic tradition calls it "metanoia", that is, a radical change of the mind, the very way of being, the symbolic death of the old (old) person, and the birth of a new person who has attained godlikeness (holiness). A similar spiritual transformation (described in terms of "enlightenment", "liberation", etc.) can also be found in Far Eastern practices.

RELIGIOUS INSPIRATION

One of the constitutive elements of religious experience is "inspiration." "Spiritual inspiration" requires an appropriate attitude, preparation, and entry into a certain psycho-emotional state, which today's psychologists would call an "altered state of consciousness", or "trans-personal experience" (K. Wilber). Such spiritual inspiration was often a source of knowledge and creativity in the lives of prominent figures.

The theme of inspiration, associated with divine madness-ecstasy, is developed in the works of Plato. This topic shows how artificial and far-fetched the modern secular division between philosophy and religion is. After all, for Plato, divine inspiration acts as the religious foundation of philosophical experience, without which we cannot leave ordinary everyday life ("the cave") and touch the foundations of the existent, super-intelligent whole. Ecstatic inspiration is a source of creative thinking, capable of imagination, the art of thinking in images, not concepts. Madness and death, are two sides of the same fundamental question about being, because we, as thinking beings, find in death the limit of our bodily (animal) existence, and in madness, we find the limit of our rational existence. At the same time, both death and madness are closely related to the essence of philosophizing. Philosophy is designed to prepare a person for death when he can finally leave the cave of ignorance and see the real world of eternal ideas.

Inspiration starts from wonder. Without it, a person will never begin to think philosophically, that is, to look into the essential depth of things, and not to slide on their surface: "For this feeling of wonder shows that you are a philosopher, since wonder is the only beginning of philosophy" (Plato, 1925. Theaetetus, 155 d). Surprise and astonishment express the appropriate mood of a person who has witnessed something extraordinary, amazing, unusual, something that goes beyond ordinary perception and common sense. Astonishment means entering the realm of the non-self-evident, which amazes and stuns us. To be very surprised means to get into a state of frenzy, esctasy (delirium), literally "to go beyond the limits of reason."

As we have already noted above, for Plato, the amazing is in a certain way correlated with the divine, the superhuman. The divine is revealed to man in ecstatic wonder-inspiration. This discovery is a gift from the "gods". "Metaphysical leap" to a new vision means momentary obsession and insane frenzy, "mania", i.e., "divine change of the usual state" (Phaedrus 265 a). Mania as a gift of God is a source of the highest benefits and true knowledge. *"Madness, which comes from god, is superior to sanity, which is of human origin"* (Plato, 1925. Phaedrus, 244 d). Prudence is a human virtue, and madness-inspiration is a God-given opportunity to see (=understand) the intelligible world of ideas.

Plato (through the mouth of Socrates) distinguishes two types of madness (frenzy): *"There are two kinds of madness, one arising from human diseases, and the other from a divine release from the customary habits"* (Plato, 1925. Phaedrus, 265 a). In the divine mania, Socrates singles out four types: the prophetic, originating from Apollo, the mysterious, inspired by Dionysus, the poetic, sent by the Muses, and the erotic, given by Aphrodite and Eros. Prophecy is a special divine gift of inspiration. Mania does not mean lack of intelligence, feeble-mindedness or short-sightedness, loss of adequacy, lack of self-awareness, etc. This is a special delight of the mind, going beyond common sense. Apollonian prophetic frenzy is "cathartic" (purifying) thinking. It can also be called morally directed and ascetic thinking, because in Socrates' prayer at the end of the dialogue "Phaedrus" the following lines are heard: *"O beloved Pan and all ye other gods of this place, grant to me that I be made beautiful in my soul within and that all external possessions be in harmony with my inner man"*. ((Plato, 1925. Phaedrus, 279 b) The combination of physical and spiritual light is the main attribute of the god Apollo. In his purity, Apollo is like sunlight, which brings bright insight, and the wild inspiration that Apollo gives is the spiritual clarity of joyful knowledge that reveals the truth. Apollo is a god of order and measure, the formative spirit. Apollo embodies man's desire for a beautiful form that combines different parts into a single organic whole.

If the epithet of Apollo is pure, radiant, and shining, then the epithet of Dionysus is the one who frees, allows, and resolves. This deity embodies the idea of becoming, movement, spontaneity, and unpredictability. Apollo separates and establishes a border and a limit, a norm that cannot be crossed. Dionysus, on the contrary, unites and abolishes boundaries, contradictions, and norms, blurs the individual, and allows stopping being oneself. It cancels meaning and order and captures the soul in the darkness of chaos - a space where opposites, life, and death converge.

Apollonian "cathartic" and ascetic and Dionysian revealing and emancipating thinking are closely related to poetic inspiration. *"For all the good epic poets utter all those fine poems not from art, but as inspired and possessed, and the good lyric poets likewise"* (Plato, 1925. Ion 533 e). The poet collects his creativity in the honey springs in the gardens and groves of the Muses. The poet can create only in a state of delirium when he allegedly loses his sense of common sense. Poetic mania involves passionate prayer, and pleading with the gods to bestow poetic inspiration.

The best type of madness is love or erotic. Eros expresses the experience of wholeness, combining spiritual and material nature in man, divine, and animal. *"All my discourse so far has been about the fourth kind of madness, which causes him to be regarded as mad, who, when he sees the beauty on earth, remembering the true beauty, feels his wings growing and longs to stretch them for an upward flight, but cannot do so, and, like a bird, gazes upward and neglects the things below"*. (Plato, 1925. Phaedrus 249 d). Erotic mania captures a person, gathers all his intellectual, volitional, and sensory-emotional forces into a single whole, raises him to the transcendental source of all things, and bestows the highest happiness-bliss.

Now let's say a few words about the symbolism of religious experience. The reality revealed in spiritual experience is higher than language. It is understood not discursively, but intuitively and contemplatively. Let's recall, how in Plato's dialogue "Cratyl" Socrates claims that language does not reach the truth of the being and that we must know the being from itself, without using names (words). Thinking, which intuitively grasps ideas, can come out of the power of names. Thus, from the point of view of this logic, the beginning and end of the movement of thought are silence, intuitive contemplation, essential complicity with things, and discourse and argumentation are just pauses in silence.

A parallel is suggested between ancient Greek and ancient Chinese doctrines, particularly, the Tao (Chapter 8). Lao Tzu wrote about the "nameless Tao" as the only secret principle, mysterious origin, and source of everything. It cannot be named, but it can be indicated. The famous Chinese scholar E. Torchinov notes: *"For the Taoist patriarch, thought and being are one, but Lao Tzu is a master of "dark things", so for him, it is not language that imposes its order on being but being itself penetrates through language... It is a language that returns to a pure message-combination. Being shines in it"* (Torchinov, 2000). It is about connectedness, complicity with the reality of what is on the other side of things, and permeates these things like blood vessels. Things are combined, interconnected, and in cooperation without any words. However, the unity of things cannot be grasped categorically. *"Whatever Lao Tzu talks about, he means "other" and even "eternal other"* (Torchinov, 2000). If the meaning lies on the other side of speech and things, then we have nothing left but to repeat after Lao Tzu: *"he who knows does not speak, he who speaks does not know"* (Torchinov, 2000). This ancient Chinese aphorism can be compared with a similar statement by Ludwig Wittgenstein. Should be kept silent does not mean that it does not exist in reality, or that it has no meaning and value. On the contrary, it is absolutely significant, but cannot be considered by positive science with its reliance on "atomic facts". Moreover, the meanings of things themselves are not "atomic facts," because they are not empirical objects. Absolutely significant, which is kept silent or symbolically hinted at, is the fact of a special unordinary experience.

SYMBOLIC DEATH AND SELF-TRANSCENDENCE.

The desire for self-transcendence, self-overcoming of one's dilapidation, one's "ego" in the direction of the Self, that is, the movement towards the true essence, towards the godlikeness of the individual, is a basic, existentially rooted characteristic of a person. However, it should be noted that all forms of manifestation of self-transcendence are connected in one way or another with death and destruction. Of course, it is not about physical death, but symbolic dying and rebirth. Bishop Callistus (Ware) in his book "The Inner Realm" notes that every significant moment of our existence is a transcendence, because *"all life is woven from a chain of little deaths and births."* (Calliste Ware, 2000). As the poet Thomas Eliot said, *"The time of death is every moment"* [Eliot, Four Quartets, 1999]. Such a "lesser death" is painful, it makes us suffer, worry, and despair, but at the same time, it can become a source of new life. "Lesser death" is associated with the fear of loss, the opportunity to discover something new for oneself. From a courageous attitude to "symbolic death" as a constructive phenomenon that should be accepted as inevitable, a corresponding attitude also follows to "real death" - the one that will happen at the end of our life. After all, if you did not regret and were not afraid to "die" (give, take risks, part with the usual, forgive, step over yourself every time, leaving the "comfort zone") during your life, then all these were preparations for the main test. In the context of Abrahamic religions (Judaism, Christianity, and Islam) every moment of our life is a rehearsal for death. The old is irrevocably leaving our lives, and the new is rushing in. We overcome the fear of this transition with faith in miraculous rebirth, renewal, and resurrection. It is noteworthy that in Far Eastern religious cults life is also considered to be a symbolic cycle of death and birth. We can recall the Buddhism of Tibet and the "bardo" state specific to this religion. In a narrow sense, bardo refers to the period between human incarnations - the state between death and a new incarnation. But in a broad sense, it is any intermediate state, any transition from one state to another. As Francesca Freemantle notes, *"Our real 'now' is a continuous bardo, an intermediate state between the past and the future"* [Fremantle, 2001]. So, bardo is a transitional state that symbolizes the beginning and the end. In this state, a person often feels fear, and uncertainty, and has to choose and make decisions. According to the "Tibetan Book of the Dead", there are six types of bardos. Having learned to control ourselves in transitional states even in this life, we will be able to learn to behave correctly even at the time of death, to curb our fear of the complete disappearance of our "I".

Symbolic death and rebirth in the process of initiation can take different forms. Mysteries are one of the most vivid examples of spiritual dedication. Speaking about the silence in which the indescribable divine reality is conveyed, and the mysterious transformation of a person through death and resurrection takes place, it is appropriate to mention the ancient Greek Eleusinian Mysteries. To this day, there is very little documentary evidence about them. This was due to the prohibition, on pain of death, to tell the participants of the ceremony what they experienced. The isolated quotes from ancient works, that hint at some details of the ceremony, have come down to us. In general terms, it is clear that at the heart of the Mysteries lay the desire of man to reunite with the deity and obtain a blessed eternal life. In the Homeric hymn in honor of the goddess Demeter, there are the following lines:

> The holy things, which it is unlawful to transgress or find out,
> Or speak of; for the great reverence of the gods checks their tongue.
> Blessed is the earth-going human who has seen these rites!
> But whoever has not carried out the holy rites, and has not shared in them,
> Will never have the same portion when he or she dies and goes under the murky gloom.
> But when the divine among goddesses had done all she had promised,
> She went back to Olympus among the assembly of the other gods.
> (475-80 lines) [Nagy, 2018].

We find similar words in the Pindar poetry: "Happy is he who saw this (mysteries) before going down to the earth. He knows the end of life. He also knows its beginning" [Pindar, 1947], and Sophocles:

> Thrice happy they, who, having seen these rites,
> Then pass to Hades: there to these alone
> Is granted life, all others evil find.
> (Fragment 719) [Sophocles, 1906].

Thus, the initiate was promised that his soul would enjoy bliss after death and it would not become a mere "shadow without memory or power" (a state so feared by all of Homer's heroes). Mysteries were supposed to remind mortals of the events of sacred history, to symbolically reproduce the plot of the abduction of Demeter's beloved daughter - the beautiful Persephone - by Hades, the ruler of the kingdom of the dead, and the subsequent happy reunion of the two goddesses. In this way, the mysteries demonstrated the erasure of the insurmountable boundary between the world of the dead and the world of the living, the possibility of symbolic mediation between the worlds.

The story of Persephone gave humanity hope for the miracle of resurrection. In this aspect, the mystery gives mortal man physical immortality. From the mythical story, we remember that Demeter wanted to turn the king's son into a god by burning him in a furnace, but by a fateful coincidence of circumstances, the people themselves prevented the completion of the matter. Demeter's failure led to the goddess's epiphany to the people and the establishment of the Mysteries. From the very moment, the divine seed is planted in the human soul and initiates the internal transformation of a beast-like being, dominated by animal instincts and passions, into a man as a god-like being, in which the spirit dominates the flesh. A demonstrative illustration of this process is Apuleius' novel "Metamorphoses, or the Golden Ass", which tells the story of the transformation of a man into a donkey due to dark magic. The donkey's many journeys have an unexpected ending: the help of the goddess Isis restores his human face. Having experienced a spiritual rebirth, the main character becomes a servant of her cult. The religious conversion of the main character symbolizes the overcoming of the crude animal nature and moral improvement through catharsis. The blows of blind fate end with the mysterious award-consecration of the main character: *"All the uninitiated were ordered to depart, I was dressed in a new-made robe of linen and the high-priest, taking me by the arm, led me into the sanctuary's innermost recess. And now, diligent reader, you are no doubt keen to know what was said next, and what was done. I'd tell you, if to tell you, were allowed; if you were allowed to hear then you might know, but ears and tongue would sin equally, the latter for its profane indiscretion, the former for their unbridled curiosity. Oh, I shall speak, since your desire to hear may be a matter of deep religious longing, and I would not torment you with further anguish, but I shall speak only of what can be revealed to the minds of the uninitiated without need for subsequent atonement, things which though you have heard them, you may well not understand. So listen, and believe in what is true. I reached the very gates of death and, treading Proserpine's threshold, passed through all the elements and returned. I have seen the sun at midnight shining brightly. I have entered the presence of the gods below and the presence of the gods above, and I have paid due reverence before them"*. [Apuleius, 1989]. In initiation, the second birth of a person takes place, which depends on the posthumous participation of the soul, or what Christian theology will later call salvation.

Rites of small mysteries were accompanied by sacrifice, fasting, and eating sacred food. They also included wanderings in the dark (symbolizing the soul wanderings after death), various terrifying spectacles, theatrical performances, and an unexpected exit to a brightly lit meadow, symbolizing the soul's liberation from suffering and its attainment of bliss. Walter Otto writes, that the ceremony participants were shown an ear of wheat, and then a "miracle" took place. The ear of corn grew and ripened with incredible speed, just as during the Dionysian Mysteries a grapevine grew in a few hours. At the ceremony climax, there was a play depicting Persephone being reunited with her mother. The neophyte joined the divine mystery and a radical consciousness change took place. *"In initiation, the neophyte felt close to the divine world, as well as the close connection of life and death. The revelation about the intersection of life and death was supposed to reconcile the neophyte with the necessity of his death"* (Eliade, 1981).

René Guénon highlighted the profound metaphysical significance of the mysteries. According to his reconstruction, the lesser mysteries related to the development of the capabilities of human nature led to perfection, the restoration of the original "heavenly" (blissful) state, and the great mysteries led to the realization of superhuman, unconditioned states - what Eastern traditions call "final liberation" or higher identification. Climbing up the steps of the states of being, a person had to become a living mirror in which the Deity contemplates his essence, and his heart is the dwelling place and temple of the Spirit. R. Guenon also draws attention to the common etymology of the words "myth" and "mystery". The word "myth" comes from the words "closed mouth", and "mute", and therefore means "silence". In turn, mystery, and secrecy are also connected with the idea of silence. The words "mystic" and "bridge" (initiate) are related, related to initiation. So, a mystery is something, that cannot be talked about, and that is forbidden to be told to outsiders. There is also a similarity between the words sacred and secret. The sacred is inaccessible to the worldly, distant, and separated from it. So, the mystery concerns what should be accepted in silence, and what should not be argued about. Guenon emphasizes that religious dogmatics are always mysterious. It contains truths that have a super-rational and super-individual nature and cannot be discussed. The mystery is vague and must be contemplated in silence, conveying the inexpressible in a symbolic form (Guenon, 2004).

RELIGIOUS INSPIRATION IN CHRISTIANITY

The Christian (primarily Byzantine) tradition also understands divine inspiration as the core of spiritual transpersonal experience and thus continues the mystical line of philosophizing begun by the Greeks. The understanding of God changes, but the awareness that in the religious experience the individual closed structure of the "ego" is opened and introduced into the sphere of the super-individual, sacred, super-essence remains unchanged. The famous Orthodox ascetic of the 20th century, St. Sophronius (Sakharov) writes about the beginning of his spiritual life, his spiritual search for God. It started from thoughts about death and the futility of human existence - thoughts that were born against the background of social disasters, wars, and cataclysms of the 20th century.

"From a young age, the thought of eternity was absorbed into my soul," writes Father Sophrony, *"If I really die, that is, sink into nothingness, then all other people like me also disappear without a trace. Therefore, all is vanity; real life is not given to us. All world events are nothing but an evil mockery of a man"* (Sakharov, 1985). The nothingness we are talking about here is not the creative "Nothingness" of God, but the physical and spiritual destruction and annihilation of the human Self in the literal sense. The author compares the death of the individual with the loss of the world: both the material universe and the inner world of man, and the Lord God: *"The suffering of my spirit was caused by external catastrophes, and I naturally*

identified them (catastrophes) with my personal fate: my dying assumed the character of the disappearance of everything I had learned, with which I was materially connected. ... My inevitable death was only something infinitely small: "one less." In me, with me, everything that was covered by my consciousness dies loved ones, their suffering and love, all historical progress, the whole Earth in general, and the sun, and stars, and boundless space; and even the Creator of the world Himself, and He dies in me; all of Being is absorbed by the darkness of oblivion" (Sakharov, 1988). This is how the young ascetic perceived his death - as the absence of being, misery, darkness, and emptiness.

But the awareness of the death tragedy, the disappearance of self-awareness in the non-being terrible gloom, instead of affirming the futility opinion and illustration of being as such, unexpectedly leads to the opposite thesis. It demonstrates the fragility and at the same time the greatness of man in the light of eternity, the self-worth of the human personality as a microcosm and its separation (exclusivity) from the rest of the created world: *"Eternal oblivion, like the fading of the light of consciousness, terrified me. This condition was destroying me; possessed me against my will. What was happening around me obsessively reminded me of the inevitability of the end of world history. ... The memory of death, gradually growing, reached such a strength that the world, our entire world, was perceived by me as a kind of mirage, always ready to disappear into the eternal lapses of nothingness. The reality of another order, unearthly, incomprehensible, took over me, despite my attempts to evade it. I remember myself perfectly: I was in everyday life, like all other people, but at times I did not feel the ground beneath me. I saw her with my eyes as usual, while in my spirit I was drowning over a bottomless abyss.... And so, without thinking about anything, a thought suddenly entered my heart: if a person can suffer so deeply, then he is great by nature. The fact that with a person`s death the whole world and even God dies is possible only because this person is in a certain sense the center of the universe. And in the eyes of God, of course, it is more precious than all other created things"* (Sakharov, 1988).

Sometimes the mortal memory took on terrible forms, in which the ascetic's contemplation of the extinction of any life manifested itself. This is how Sofroniy Sakharov describes one of his inner spiritual experiences: *"I vividly remember one of the most characteristic days of those days. I read while sitting at the table; I support my head with my hand, and suddenly I feel a skull in my hand, and I mentally look at it from the outside"* (ibid.). The memory of death conditioned all further ethics of attitude toward the Other and built a kind of hierarchy of values, where the eternal was more important than the temporary: *"Everything that was subject to decay was devalued for me. When I looked at people, before any thought I saw them in the power of death, and my heart was filled with compassion for them"*. (ibid.). Such sympathy-empathy made impossible both a pragmatic-utilitarian approach to *one's neighbor and symmetry: "I wanted neither glory from the "dead" nor power over them; I didn't expect them to love me. I despised material wealth and did not highly value intellectual wealth, which did not give me the answer to what I was looking for. If I were offered a lifetime of happiness, I wouldn't take it. My spirit needed eternity, and eternity, as I later realized, stood before me, truly rebirthing me. I was blind, out of my mind. The eternity was knocking at my soul`s door, that was locked in fear to itself"* (Sakharov, 1988).

It is the memory of death, according to Sophrony Sakharov, that marks the beginning of salvation and the spiritual struggle against sin. *"The memory of death is a special state of our spirit, not at all similar to the knowledge inherent in all of us that one day we will die. This wonderful memory takes our spirit out of earthly gravity, frees us from the power of temporary passions and preferences, and thus makes us live naturally holy. Although in a negative form, it nevertheless holds us tightly to the Eternal"* (Sakharov, 1988).

So, the memory of death gives the ascetic some previous experience of dispassion. It stops the action of passions and thereby lays the beginning of a fundamental change in our entire life activity and the nature of perception of all things. The fact that it allows us to experience our death as the end of the entire universe confirms the revelation given in the Bible that man is the image of God, and as such he can contain both God and the created cosmos. Through awareness of one's own mortality and finitude, man realizes the hypostasis principle. The experience of death memory prepares a person for a more realistic perception of the Christian Revelation and that theology, which is based on the experience of a transcendental being.

However, Sophrony Sakharov further notes that the greatness and spiritual height, originally given to man at the creation of the world, with the correct passage of spiritual life, are combined with the awareness of the loss of this godlikeness due to the fall of man. As the Absolute Being is revealed to us in the prayer experience, we increasingly feel our worthlessness and impurity, our sinfulness. Yes, man is great as the image of God, but he lost his greatness by damaging his nature, and now he must restore this greatness and heal his damaged nature by turning to Christ the Savior for help. God inspires man. *"With all the awareness of my extreme worthlessness, I prayed for decades that the Lord would give inspiration to follow Him wherever He went ... By inspiration I understood the presence of the power of the Holy Spirit within us. This kind of inspiration belongs to a different order of being so it is not artistic or philosophical inspiration. It can be understood as a gift from God, but it still does not give either a personal union with God or even intellectual guidance (understanding) about Him. Truly holy inspiration, which comes from the Father, is not imposed by force on anyone: it is received, like any other gift from God, by intense effort in prayer. ... All of us need to undergo a complete rebirth by the action of grace, restoration of our ability to perceive deification"* [ibid]. Following the *tradition of the holy fathers, the ascetic calls this spiritual rebirth under the influence of inspiration deification (theosis). It consists in overcoming spiritual death and obtaining eternal life. Moreover, deification is a co-action (synergy) of the individual Self and the divine Other. Only thanks to these joint efforts, in submitting one's will to the divine Revelation, a person can live a righteous life according to the Gospel commandments, among which the central place is occupied by love-mercy. The Gospel commandments are called "commandments of beatitudes", accordingly, living according to these commandments gives a person the opportunity to feel the grace, the beatitude, or "the peace of Christ": "True deification consists in the fact that the infinite life of God himself is really and forever transmitted to a rational being ... Inspiration from above depends on us: will we open the door of our heart so that the Lord - the Holy Spirit, who stands at the door of our heart and knocks - does not forcibly enter us. When the soul (exactly in this way, i.e., through divine inspiration) touches this Eternity, then the base passions fall away from us ... the "peace of Christ" descends upon us, and we gain the power to love our enemies and carry out other commandments of Christ"* (Sakharov, 1988).

Another Orthodox ascetic of the 20th century, the abbot of the Valaam monastery, Khariton, also writes that a person's prayerful presence before God, his authentic, not imitative religious experience, involves "synergy", that is, the joint action of human and divine will, a combination of free will and involuntary. "What are they looking for with the Jesus prayer? - So that a blessed fire sinks into the heart, and continuous prayer begins, which is exactly what determines the state of grace ... The prayer of Jesus, when the God spark falls into the heart, blows it into a flame, but it does not itself give this spark but only promotes its acceptance. The spark of God is a ray of grace. You cannot cause it by anything, it comes directly from God ... This fire is not attracted by any skill, but is given freely by the grace of God. If you want to receive prayer, work in prayer. God, who sees how you diligently seek prayer, will give you prayer Himself" [Hariton, 1936]. The nature of religious transpersonal experience is such that it certifies and verifies itself. Here we agree with W. James, who emphasized that mystical states that have reached their full development are absolutely authoritative for the persons who experience them. This can be compared with the statement of the famous Orthodox ascetic Siluan of Athos, who notes: *"When God presents himself in a great light, then there can be no doubt that he is the Lord, the Creator and the Almighty"* (Elder Sophrony, 1997).

The list of prominent individuals who experienced religious inspiration could go on, but other important questions arise: to what extent can a transpersonal state of consciousness be induced artificially; where the possibilities of human will end, and superhuman forces come into play, the ontological character of which cannot be doubted. Awareness of human weakness and finiteness, non-autonomy is the main prerequisite for the transpersonal experience. So, firstly, not everything depends solely on a person's active efforts, his knowledge, and the practice of the relevant methods and techniques of induction of experience. In this regard, the experience is mostly passive, when a state of ecstasy-frenzy and grace "descends" (without warning), and a man plunges into it without his own will. Secondly, in this experience passivity, the ontological (real or essential) character of the transcendent force that causes it is experienced. Thirdly, only in a supernatural state of grace, a person is able to put into practice the theological virtues of faith, hope, and love.

SUBSTITUTES OF SPIRITUAL EXPERIENCE.

Some youth subcultures that appealed to transpersonal experience and metaphysical nostalgia in the secular world should be mentioned. The phenomenon of hippie culture with its sexually unbridled, black rhythms of rock and roll, and admiration for the art of the East is of great interest to researchers. The practice of using psychedelic substances in the hippie milieu (and later in other subcultures) was an attempt to escape the atomized individual beyond the usual meaningless and absurd pragmatic existence - into the supra-individual sphere. One can call it escapism - a kind of departure from social reality, caused by fear of its challenges, into some alternative artificial and illusory world.

Luigi Zoja notes in his book "Drugs, Addiction, and Initiation: Modern Search for Ritual" that these were initiation practices in modern culture, where institutionalized initiation has disappeared. At the same time, they can be interpreted as a manifestation of nostalgia for meaning, for God, or simply for something sacred, and absolutely authoritative. Metaphysical nostalgia, which is the essence of the youth subculture, is not a memory of what once happened and will be experienced in ordinary life experience, but an expression of the transcendental experience of eternity, immortality, and heavenly bliss. Modern society lacks initiatory rites that used to signify the symbolic death and transformation of the individual, the transition from nature to culture, from the profane to the sacred. *"There is a latent need for initiation in our society, and this leads us to suspect that the mass drug use that has occurred since the late 1960s (as well as the initiation of esoteric groups) is a chaotic expression of this need. Our society lacks initiation rituals and death rituals because it is a forbidden topic in society"* (Zoja, 2000). At the moment of taking drugs, a person experiences a more or less intense death, he distances himself from rationality, consciousness, and lucidity, to which we are all doomed by the main imperative of Western civilization. Drugs are associated with rebellion, *"the denial of mainstream Western cultural and ideological dominance. This dominance includes the primacy of the rational over the irrational"* (Zoja, 2000).

The constant growth of the practice of using narcotic substances, especially in the developed countries of the West, is a rather revealing moment. The Italian conservative thinker J. Evola in his work "Ride the Tiger" (Evola, 2003) wrote that in the 1950s and 1960s drugs began to be perceived by the "advanced informal youth" as vital food, an answer to existential anxiety, protection from the abyss of Nothingness, various social fears-phobias, and the growing fear of death. *"First of all, we are talking about young people who more or less acutely experience the emptiness of modern existence and the routine of the current civilization and strive to avoid it"* (Evola, 2003). This is the original reason, although more prosaic factors (fashion, conformism) are added later. The problem with drugs, from Y. Evola's point of view, is that here *"means, which were originally used as aids for penetrating into the supersensible, into the area of initiatory or similar experiences, are transferred to the profane and physical level"* (Evola, 2003). Modern drugs are similar to those narcotic substances that primitive peoples took for sacred purposes, within the limits of the Tradition (for instance, the use of narcotic substances by shamans). Drugs allow an individual to forget, *"to get out of himself, to passively open himself up to states that give the illusion of higher freedom, intoxication, and extremely sharpen sensations"* (Evola, 2003). But in fact, such an influence has a destructive character on the personality. It is known that in ancient times, shamans took psychoactive substances (mescaline, ayahuasca) to contact the world of spirits, and modern man has a single goal - to get rid of tension, trauma, stress, neurosis, a feeling of emptiness and the absurdity of existence, or even physical pain in the case of serious diseases in the terminal stage. The ecstatic nature of the narcotic dope opens the way to the invasion of dark, demonic forces that dominate a person.

Y. Evola emphasizes that some narcotic substances, such as mescaline, cause an aggravation of all sensations, a vision of an artificial paradise, full of illusory phantasmagorias, which *"are not able to open the door to the real transcendental realm"* (Evola, 2003). However, it cannot be excluded that by taking drugs, a person opens access to dark demonic influences, that is, instead of initiation as death and rebirth in a new, higher superhuman quality, he does, without realizing it, a counter-initiation, an ontological descent down into the sub-human. This is confirmed, in particular, by the fact that in a narcotic state, a person commits completely immoral and illegal acts, for example, rape, theft, robbery, murder, or suicide, and loses self-control. Therefore, all this can be called a surrogate of spiritual experience, a counter-initiation, the result of which is not a renewal of life in the spiritual plane, but a meaningless spiritual and social degradation and, ultimately, death without hope of resurrection.

In today's postmodern realities, a person is characterized by a lack of a spiritual component, closedness, and isolation in his existence, in which the supremacy of material values is asserted and things of a transcendental order are ignored. Negative phenomena of loneliness, existential vacuum, depression, suicidal moods, and unrestrained hedonism are symptoms of this spiritual decline. The way out of the existential crisis is the "turn of consciousness", the existential attitude of the individual to the sacred secret of being as a higher spiritual authority, which is capable of filling human existence with meaning, significance, and happiness (Morozov, 2017). It takes away the fear of death and reveals the mystery of personal immortality. The authentic existence of an individual implies openness and orientation towards the realm of eternal "supreme values". The highest manifestation of spirituality is a trans-personal experience, which is constitutive in the process of humanization, bringing a person closer to his essence. The trans-personal experience in which a person experiences the presence of divine reality as unconditional love and the source of all good is what opens the way for the individual to see others in the light of this love as unique and inimitable You. Through love- mercy, an individual is assimilated to God, and a spiritual, hypostatic (personal) component appears in his biological nature. Among the features of transpersonal experience is its ambivalent active-passive character, where transcendence, that is, outgrowing one's limits of individual consciousness, is never pure self-transcendence. It is emphasized that the reality of the absolute, which is revealed in experience, is understood not rationally-discursively, but intuitively-contemplatively, through essential complicity. In religious inspiration, a person discovers the indescribable mystery of the divine and the richness of its meanings, a qualitative initiatory transformation through symbolic death and resurrection.

It depends on a person's will of whether one wants to give up his nihilistic and cynical utilitarian-pragmatic outlook on the world and change one`s mode of being to a useless "substantial" attitude to reality. Nostalgia for the fullness of being, for a complete picture of the world, on the one hand, and the soulless technocratic and object character of modern civilization, in which alienation and impersonality prevail, on the other hand, should become a turning point on the way to the main event of history - the revenge of the sacred.

CONCLUSION

The problem of a person's finiteness is inextricably linked with the state of his consciousness (both the "spirit of the times" and the individual) and the corresponding definition of the meaning of existence.

In today's postmodern realities, ordinary human consciousness is characterized by a lack of a spiritual component, isolation, isolation in its existence, the dominance of material values, and a complete disregard for things of a transcendent order.

The symptoms of a lack of spirituality are the phenomena of loneliness, existential vacuum, depression, suicidal moods, unbridled hedonism, etc.

The way out of the existential crisis is a "turn of consciousness" towards the sacred mystery of being as the highest spiritual authority that can fill a person's existence with meaning, significance, and happiness. The authentic being of a person presupposes openness and orientation to the area of eternal "supreme values". At the same time, the highest manifestation of spirituality is transpersonal experience, which contributes to the comprehension of one's essence through symbolic death and rebirth. Transpersonal experience, in which a person experiences the presence of divine reality as unconditional love and the source of all blessings, opens the way for the individual to see others in the light of this love as unique and unrepeatable entities. Through love-mercy, an individual is likened to God, and in his biological nature, a spiritual, hypostatic (personal) component emerges.

Among the transpersonal experience features, its ambivalent active-passive character is indicated, where transcendence, that is, the outgrowth of one's boundaries of individual consciousness, is never pure self-transcendence. It is emphasized that the reality of the absolute, revealed in experience, is comprehended not rationally-discursively, but intuitively, through existential involvement.

Examples of such personal transformation in the ancient world are the Eleusinian Mysteries, and in Christian times, the sacraments of the Church, especially the sacrament of the Eucharist. Pre-Christian and Christian mysteries-mysteries give a person hope for resurrection, immortality, and eternal bliss in eternal life. The symbolism of the sacraments is not pictorial, but epiphanic, where the spiritual reality is revealed and manifested in material signs, uniting the intra-world and the extra-world.

According to Heidegger, the roots of the modern "disenchantment of the world", nihilism, and oblivion of the sacred truth of being should be sought in objective onto history. It depends on the free will of a person whether he wants to abandon the nihilistic and cynical utilitarian-pragmatic attitude and change the mode of being to a disinterested "essential" attitude towards reality. Nostalgia for the fullness of being and a holistic picture of the world, as opposed to the soulless technocratic nature of modern civilization, should become the driving force on the way to the main Event of onto history - the revenge of the sacred and the return of the "last god".

REFERENCES

Apuleius, L. (1989). *Metamorphoses (The Golden Ass)*. Harvard University Press.

Baloyi, L., & Makobe-Rabothata, M. (2014). The African conception of death: A cultural implication. *Toward Sustainable Development through Nurturing Diversity*. https://doi.org/DOI: 10.4087/FRDW2511

Bedau, M. (2014). *The Nature of Life*. Cambridge University Press.

Behrendt, K. (2018). Unmoored: Mortal harm and mortal fear. *Philosophical Papers*, 48(2), 179–209. DOI: 10.1080/05568641.2018.1462668

Charlier, P., & Deo, S. (2018). The notion of soul and its implications on medical biology. *Ethics, Medicine, and Public Health*, 5, 125–127. DOI: 10.1016/j.jemep.2018.05.005

Cholbi, M. (2016). *Immortality and the Philosophy of Death*. Rowman & Littlefield.

Cholbi, M. (2022). *Grief: A Philosophical Guide*. Princeton University Press. DOI: 10.2307/j.ctv1n1bs19

Cholbi, M. (2023). *Oxford Handbook of the Philosophy of Suicide*. Oxford University Press.

Dasgupta, S. A. (2014). *History of Indian Philosophy*. (5 vol.). Cambridge University Press. (Original work published 1922)

Egerstrom, K. (2021). *Making death not quite as bad for the one who dies*. Routledge.

Eliade, M. A (1981-1988). *History of Religious Ideas*. (3 vol.). University of Chicago Press.

Eliot, T. Four quartets. *T.S. Eliot: Four Quartets - an accurate online text*. (n.d.). Davidgorman.com. Retrieved August 15, 2024, from http://www.davidgorman.com/4quartets/

Evola, J. (2003). *Ride the Tiger: A Survival Manual for the Aristocrats of the Soul*. Inner Traditions/Bear.

Fischer, J. M. (2019). *Death, immortality, and meaning in life*. Oxford University Press.

Frankl, V. (2006). *Man's Search for Meaning*. Beacon Press.

Fremantle, F. (2001). *Luminous Emptiness Understanding the Tibetan Book of the Dead*. Shambhala Publications.

Gadamer, H.-G. (1989). *Truth and Method* (2nd ed.). Continuum Intl Pub Group.

Guénon, R. (2004). *Perspectives on initiation*. Sophia Perennis.

Hagglund, M. (2019). *This Life: Why Mortality Makes Us Free*. Profile Books.

Kamm, F. M. (2021). *Almost over: aging, dying, dead*. Oxford University Press.

Kaufman, F. (2016). *Lucretius and the Fear of Death*. Cholbi.

Kaufman, F. (2022). Death, deprivation, and a Sartrean account of horror. *Pacific Philosophical Quarterly*, 103(2), 335–349. DOI: 10.1111/papq.12353

Kokosalakis, N. (2020). Reflections on Death. *Philosophical/Existential Context. Society*, 57(4), 402–409. DOI: 10.1007/s12115-020-00503-5 PMID: 32836561

Manabu, F. (2023). Exploring the philosophical concept of my death in the context of biology: The scholarly significance of the unknown. *Continental Philosophy Review*, 56(2), 317–333. DOI: 10.1007/s11007-022-09596-7

Maslow, A. H. (1968). *Toward a psychology of being*. Van Nostrand.

May, R. (1996). *The Meaning of Anxiety*. Norton.

Mercer, S. A. B. (2020). *The Pyramid Texts*. Global Grey. (Original work published 1952)

Morozov A. (2017). Moral and Religious Motives in the Works of J.R.R. Tolkien: Cultural Context. *Ukrainian cultural studies, 1*, 60-65

Morozov, A., Nikulchev, & M., Kulagin, Y. (2021). Religious experience and temptations of nihilism in spiritual life of personality. *Beytulhikme. The international journal of philosophy, 11*(2), 683-697.

Morozov A. (2022). Death as existential problem: philosophical and religious aspects. *The journal of oriental studies. Institute of oriental philosophy* (IOP), *31*, 172-191

Morozov A. (2024). Ontological dimensions of the main ethical categories: freedom, conscience, equality and dignity. *The journal of oriental studies. Institute of oriental philosophy (IOP), 33*, 251-273

Mwania, P. (2016). Interface between African's Concept of Death and Afterlife and the Biblical Tradition and Christianity. (N.d.). Tangaza.Ac.Ke. Retrieved August 14, 2024, from https://repository.tangaza.ac.ke/server/api/core/bitstreams/2ad4ce8c-5d88-496a-8986-2b23b084f309/content

Otto, W. F. (1954). *The Homeric Gods*. Thames & Hudson. (Original work published 1929)

Otto, W. F. (1995). *Dionysos. Mythos und Kultus*. Indiana University Press. (Original work published 1933)

Otto, W. F. (2015). *Teofania*. Sexto Piso Espana S L. (Original work published 1956)

Pindar, . (1947). *The extent odes of Pindar*. The University of Chicago.

Plato, . (1925). *Plato in Twelve Volumes*. Harvard University Press.

Radhakrishnan, S. (2009). *Indian Philosophy* (Vols. 1–2). Oxford University Press. (Original work published 1923)

Sakharov, S. (1988). *We Shall See Him as He Is*. Stravropegic Monastery of St. John the Baptist.

Silouan the Athonite. (1991). *The Life and Teachings*. Palomnik. Hegel, G. (1977). *Phenomenology of spirit.* Oxford university press Heidegger, M. (2006). *Sein und Zeit*. Niemeyer, Tübingen. (Original work published 1927) Nagy. Gregory. (2018). *Homeric hymn to Demeter*. University of Houston Press. Kovalyov, Y., Mkhitaryan, N., & Nitsyn, A. (2020). *Self-organization of the Human Mind and the Transition from Paleolithic to Behavioral Modernity.* IGI Global International Publisher of Progressive Information Science and Technology Research.

Sophocles, . (1906). Tragedies and fragments. *Health*.

Taylor, J. S. (2014). *The Metaphysics and Ethics of Death*. Oxford University Press.

Tillich, P. (2000). *The courage to be*. Yale University Press.

Van Dijk, T. A. (2006). Discourse, context, and cognition. *Discourse Studies*, 8(1), 159–177. Advance online publication. DOI: 10.1177/1461445606059565

Wallis Budge, E. A. (1967). *The Book of the Dead. The Papyrus of Ani*. Dover Publications. (Original work published 1895)

Callistos Ware. (2000). *The Inner Kingdom: Collected Works*. ladimir's Seminary Press

Wright, M. R. (1985). *The presocratics: The main fragments in Greek with introduction, commentary and appendix containing text and translation of Aristotle on Presocratics*. Bristol Classical Press.

Yourgrau, P. (2019). *Death and Nonexistence*. Oxford University Press. DOI: 10.1093/oso/9780190247478.001.0001

Zoja, L. (2000). *Drugs, addiction, and initiation: modern search for ritual*. Daimon Verlag.

ADDITION READING:

Bakhtin, M. M. (1979). *Estetyka slovesnogho tvorchestva*. [Aesthetics of verbal creativity]. Art.

Fideler, D. (2005). *Ysus Khrystos, Syn Bozhyj. Drevnjaja kosmologhyja y rannekhrystyanskyj symvolyzm*. [Jesus Christ, Son of God. Ancient Cosmology and Early Christian Symbolism]. Izdatel'skij dom Sofija. Florensky, P. (1993). *Iconostas*. [Iconostasis]. Mifril. Russkaja kniga. Hegel, G. (n / d). *Aforyzmy*. [Aphorisms]. *Literacy Research: Theory, Method, and Practice*.

Heidegger, M. (2012). *K chemu poety?* [Why poets?]. Proza.Ru. Retrieved August 15, 2024, from http://www.proza.ru/2012/02/23/632

James, W. (1993). *Raznovydnosty relyghyoznogho opyta* [The Varieties of Religious Experience]. Nauka.

Korshunov, A. M. (1991). Kateghoryja poznavateljnogho obraza. [Category of cognitive image] In *Theory of Knowledge. Socio-cultural nature of knowledge*. Nauka.

Losev, A. F. (1982). *Znak. Symvol. Myf* [Sign. Symbol. Myth]. Publishing House of Moscow University.

Losev, A. F. (1993). *Fylosofyja ymeny*. [Name philosophy]. In *Being. Name. Space*. Thought.

Lotman, Y. M. (2000). *Semiosfera* [Semiosphere]. Art-SPB.

Minin, P. (2003). *Misticizm i ego priroda* [Mysticism and its essence]. Prolog.

Morozov, A. (2009). *Ljubov i smertj: ekzystencijni aspekty* [Love and death: existential aspects]. Slovo.

Morozov, A. (2018). *Zlo: metafizychni i bohoslovs ki vymiry* [Evil: metaphysical and theological dimensions]. Kyiv National University of Trade and Economics.

Pashhenkom, V. I., & Pashhenko, N. I. (2001). *Antychna literatura: pidruchnyk dlja studentiv vyshhyh navchal'nyh zakladiv* [Antique literature: a textbook for university students]. Lybid.

Shmeman, A. (2006). *Evharistija. Tainstvo Carstva*. [Eucharist. Sacrament of the Kingdom]. Granat. Sokuler, Z. (n / d). *Ljudvig Vitgenshtejn i ego mesto v filosofii XX veka* [Ludwig Wittgenstein and his place in philosophy of the 20th century]. (N.d.). Bim-Bad.Ru. Retrieved August 15, 2024, from http://www.bim-bad.ru/docs/sokuler_vittgenstein.pdf

Sophrony, E. (1997). *Staretz Siluan 1866-1938* [The Monk of Mount Athos: Staretz Silouan 1866–1938]. St. Vladimir's Seminary Press.

Taylor, Ch. (2018). *Svetskyj vek*. [A Secular Age]. Duh, i Litera.

Torchinov, E. (2015). *Religii mira. Opyt zapredel'nogo*. [Religions of the World. The experience of Transcendent]. Akademicheskaja kniga. Hariton, the abbot of Valaam monastery. (1936). *Umnoe delanie. O molitve Iisusovoj. Sbornik pouchenij Svjatyh Otcov i opytnyh eja dejatelej*. [Clever doing. On the Jesus Prayer. Collection of the Holy Fathers' teachings and their experience actions]. Izdatelstvo Valaamskogo monastyrja.

Wilber, K. (2008). *Vseobschee videnie* [The Integral Vision]. Zolotaja Amphora.

Wittgenstein, L. (1995). *Loghyko-fylosofskyj traktat; Fylosofskye yssledovanyja* [Tractatus Logico-Philosophicus; Philosophical Investigations]. Osnovy.

Conclusion

Let us summarize the main results.

1. Understanding the structure and life cycle of a person within the framework of the relationship of waves and solitons and external interactions according to the scenario (1S, 1O) allows:
   a. To model death as a process of destruction of the soliton-wave system.
   b. To propose scenarios for the post-mortem existence of the human body and sensory space as a continuation of the aging scenario under conditions of broken integrity of the soliton-wave system.
   c. To substantiate verifiable assumptions about the structures and processes of such existence.
2. A method for determining invariants and calibrations of texts and artifacts from different eras based on the apparatus of the ternary connective is proposed. This method was used to test posthumous scenarios of existence based on analyzing pre-ranked sources reflecting universal human experience. It has an independent significance in tasks of hermeneutic analysis (Kovalyov, 2023).
3. The check showed the presence of common invariants both for the proposed post-mortem scenarios and for the considered sources. It should be noted that the conclusions are valid in the "optics" set out in Chapter 2, and to the extent that the models accurately reflect the properties of the object, and the sources are reliable.

Second, consider the sequence of obtaining these results chapter by chapter.

Chapter 1, *Self-Organization* was written for the convenience of readers. The *Life Cycle in the Natural Sciences as a Complex System Self-Organization* showed the applicability of various models based on self-organization scenarios. This gives reason to prolong these scenarios for the time after death. So, we begin with a brief exposition of the axiomatics of the wave model of S-space and the theory of complex systems self-organization, models of the life cycle, and the process of dying.

Chapter 2, *Post-Mortem Models and Scenarios,* substantiates scenarios for these components' existence after a person's death. Scenarios are different:

- The soliton existence associated with the body goes through further decay (but it can be slowed down if the exchange of substances with the environment is established or special measures are taken).
- The existence of a wave associated with a person's subjective space (the definition of this space is given in (Kovalyov et al., 2020) and refined in Chapter 8) can be described according to the convolution scenario.

This convolution, called the general post-mortem scenario, its stages, and special cases, are described in detail and summarized in a table convenient for further verification. The question is natural: how correct is such an extrapolation, because medical devices do not record any oscillatory phenomena or electromagnetic fields after death, which is a well-known fact. One of the answers lies in the following reasoning: the wave model of self-organization of complex systems was subjected to a comprehensive check in the previous chapters (the conditions for the emergence of life, abiogenic and biogenic synthesis, the evolution of living organisms, including the processes of reproduction and inheritance, the structure of a person, his life cycle and interactions with environment), and the known facts did not contradict the provisions and predictions of the theory and simulation data. Then why might this theory become inapplicable after death? There are no objective reasons for this: the human body and its subjective space remain complex open systems interacting with the environment, inheriting the trends of a common life cycle, and even preserving some symmetries. And if this is so, then the laws of self-organization, expressed by various scenarios, continue to operate. However, another answer is also possible, which consists of the organization of verification - this is an analysis of universal human experience, recorded in numerous artifacts, religious texts, myths, and fairy tales. The sources of such knowledge are not considered scientific, so we are talking about only indirect verification. But there is no other way yet.

The next chapters are devoted to such verification.

The verification method: its components, sequence, methods of working with sources related to different prehistoric and historical times, and the rating evaluation of sources according to 7 proposed criteria was substantiated (Chapter 3, *Post-Mortem Scenarios Verification Method*).

The first stage of verification was carried out, combined with testing the methodology: the levels and channels of the subjective space of a person presented in Chapter 8 of the *Life Cycle in the Natural Sciences as a Complex System Self-Organization* were compared with the concepts of "spirit", "soul", Ka, Ba, kosha, and so on, used in the sources, to identify them. structure and meaning. At the same time, the structure of the subjective space also acted as a kind of "coordinate system" that made it possible to compare the data of sources with each other. The test showed the methodology's effectiveness and made it possible to determine the structure and meaning of the abovementioned concepts (Chapter 4, *Structures Checking*).

The data of sources about the events occurring for the "soul" (and other structures) after the death of the body were compared with the general post-mortem scenario substantiated in Chapter 3 and its special cases. Checking again showed a good fit - most sources describe special cases, but there are also descriptions of the general scenario (Chapter 5, *Events Checking. What Old Artifacts and Middle Eastern Texts Can Tell,* and Chapter 6, *Events Checking. Indo-Tibetan Version*).

Chapter 7, *Details from Low-ranked Sources*, based on low-ranking sources, discusses some interesting ideas about the World of the Dead - entry points, paths, "topography", inhabitants, and the fight against its masters. This chapter, of course, is not scientific and is intended to revitalize the dry text of the previous chapters with artistic details.

Chapter 8, *The Transpersonal Experience in Religious and Cultural Practices as a Response to the Challenges of Death,* presents a philosophical reflection on the inevitable individual finitude. We single out the following: individual finiteness is considered in the context of a more general problem - the meaning of human existence. The existence of modern humans has lost its sacred depth and has become meaningless; surrogates for meaning are social status, consumerism, and unbridled hedonism. The topic of death is taboo. From this point of view, a person is threatened with a complete loss of subjectivity: propaganda manipulations and pharmacological means are already capable of modulating any emotional reaction that will be indistinguishable for a person from his own. The way out is seen in the return of sacred depth and the corresponding system of values, which requires initiatory practices, important components of which are transpersonal experience and the event of death.

The philosophical view is in "another plane". In addition, results, based, on the one hand, on the use of mathematical apparatus, and, on the other hand, on sources using non-scientific methods of obtaining knowledge, naturally evoke a reaction of skepticism, which was noted in the main text. This opens up space for dialogue, which is necessary for the results' better understanding. Y. Kovalyov represents the view of a mathematician (M.), and A. Morozov - a philosopher (Ph).

M.: I would like to discuss using the apparatus of ternary connectives, invariants, and calibrations (Chapter 1) - this will allow us to establish what is common in mathematical and philosophical understanding, despite their external differences. The first thing to be stated is the awareness of the systemic nature of the problem: life and death are interrelated, and their study should consider both interactions with the environment and the history of evolution. At the same time, mathematics operates models, structures, and processes based on natural science data, while philosophy focuses more on social, moral, and religious aspects. Do you agree to discuss that?

Ph.: Yes. Philosophy should be considered as an integral worldview knowledge, which, without excluding the data of other sciences, allows you to build a complete picture of the world and fit a person into it. The subject of philosophy is the reality totality, the whole world, in other words, "everything" as unity in diversity. This "everything" includes reality and possibility, life and death. Naturally, the idea suggests expressing this unity and plurality through number and numerical proportions. So did Pythagoras, the first to call himself a philosopher, and later Plato and Plotinus, Descartes, and Leibniz. Recall that in the Platonic tradition, the beginning of everything is One (One). Being is a Unit that breaks up into a plurality, and then must again turn into the original Unit. We see that the First Cause is expressed mathematically. However, this same Unit is also called the Highest Good. This means that the roots of ethics also lie in mathematics, and we cannot do without "number" and "measure", because quantitative relations and ratios inevitably turn into qualitative ones.

In Judaic and Christian theology, the creation of the world by God comes from "nothing", here the concept of "non-existence", and "zero" already arises, which is also very interesting for the history of mathematics. After all, the Greeks and Romans, as we remember, did not have "zero" as a concept, since non-existence does not exist, just as there was no concept of "actual infinity". Having arisen from nothing, called ("hailed") to be by the divine Word, man can no longer die as an ontological entity. He is doomed to eternity. But he can die spiritually. In this sense, one should speak either about approaching the absolute spiritual life (God) or about moving away from the spiritual life towards absolute death – where there is no God.

A dialogue between philosophy and mathematics is possible when we talk about the metaphysical nature of reality, the origins of the world, the emergence of life, and the emergence of man. Let's take, for example, the anthropic principle of cosmology, according to which the world, thanks to the constants of nature, develops according to such a scenario that an observer (man) inevitably appears. The question arises, why are these constants exactly like this, and not others? How did this fine-tuning or adjustment of the material world to suit a person come about? Is it spontaneous self-organization of matter? Or, as the ancients said, the whole point is that God is a great geometer, endowing the world with measure and number? Does evolution have a reasonable purpose and meaning? Having raised questions about the purpose and meaning of evolution, the mathematician is forced to stop, go beyond the competence of his science, and step into the realm of philosophical or theological thought. Philosophy and mathematics complement each other here.

M.: Discuss the correlation of verification methods in philosophy, mathematics, and natural sciences. At one time, Aristotle created logic, without which there would be no Euclid's geometry and mathematics in general. He also wrote "On Sophistic Refutations" (Aristotle, 1978) to eliminate erroneous conclusions. Everyone remembers the contribution of Leibniz to the creation of differential calculus. The works of Bergson and Lossky on intuition have been published for quite a long time, however, intuitionistic mathematics has developed based on its premises, and only now are their results in demand (Chapter 3). Gödel's work on the incompleteness of axiomatic systems and Tarski's work on the non-coincidence of the classes of true and provable statements, in turn, did not significantly affect the methods of philosophy. Also in the natural sciences, in addition to experiments, irreproducible unique observations have long been used, as well as, for example, the principles of Bohr's complementarity, Heisenberg's uncertainty, Poincaré-Einstein's relativity, the anthropic principle, the concept of a paradigm (according to Kuhn), and so on. We see a methodological gap between mathematics, philosophy, and the natural sciences. How do modern philosophers evaluate all this? Are they involved in the development of scientific verification methods?

Ph.: It should be noted that philosophy considers the verification procedure itself as one of the ways to test the truth of theoretical statements (hypotheses, theories) by comparing their content with the content of empirical data obtained as a result of experience. The empirical tradition, begun by F. Bacon and continued by positivists and neo-positivists, dominated philosophy for a long time. One can recall L. Wittgenstein and other thinkers of this direction, who say that any scientific statement should be checked for meaningfulness. According to their system, only those scientific propositions are meaningful (and therefore capable of being true or false) and follow from sensory data and observations accessible to experiment and measurement. "The world is not a collection of things, but facts," says Wittgenstein. We know nothing about things in themselves, for us they are at best unknowable entities, "noumena" (according to Kant). We don't even know if they exist. However, it is not all so simple. If we consistently apply this methodological principle, it will turn out that scientific statements of a universal nature, among which the laws of nature belong, will be meaningless. No universal, objective, and necessary law of nature can be deduced inductively from a finite set of facts obtained in observation.

There were other philosophical attempts (for example, K. Popper) to revise the verification principle, replacing it with a falsification procedure. But all this leads away from the problem. It should be noted that philosophy does not abandon the postulate that "practice is the criterion of truth." However, this postulate should be properly understood. You quite rightly recalled the intuitionism of A. Bergson and N. Lossky, since the concept of experience is being revised in non-classical philosophy. For some reason, we reduced all the wealth of experience to a scientific experiment.

But the experience may be different. If life is an element, an endless stream, a tape, then it cannot be understood by cutting it into pieces and making "freeze frames" from concepts. Rather, this flow should be experienced intuitively, merging with it. In this merging, the bifurcation between subject and object disappears. I like the round dance example. Is it possible to understand what a round dance is outside of it? No, only by holding hands and becoming part of the dance, starting to move in a circle, you can understand it. But who is the object and who is the subject? There is a flow of life and its knowledge as an experience of unity with the whole. But, let's say, aesthetic experience, moral experience, or the experience of holiness. As S. Frank writes, all these are manifestations of a special spiritual, supersensory experience. I would add that this experience is purely human because only a person can "see" the elusive beauty (be it nature or art) not in a simple sum of contemplated things, sounds, colors, and materials, but in addition to them, through them, as a super-principle some form that unites them. Where, for example, is the beauty of Michelangelo's sculptural composition "Pieta"? In the material, color, curves, and lines of marble? Where, in which of the notes and at what moment in time, in what musical theme, laid out in the score and played by the whole orchestra, is the beauty of J. S. Bach's melodies hidden? This moment of spiritual experience is in some way self-verifying. Indeed, when the beauty of the works of Michelangelo and Bach was revealed to you, it already exists as objectively significant, truly existing. You didn't just think it was beautiful. You realize that beauty is real, not imaginary. The same applies to ethical values (love, kindness, compassion, forgiveness, gratitude) or religious experiences of sin or holiness. Here we are also dealing with facts, but facts of a special kind, facts of your inner reality, which is no less real than scientific reality. In the experience of empathy, I feel the pain and suffering of a stranger as my own, I share joy and love with him. From this point of view, the statement "I love you" is meaningful and true as soon as it is co-validated in the personal human experience of the two lovers. Another thing is that this spiritual experience, unlike the scientific one, cannot be artificially evoked or identically reproduced. After all, with all the universality and all-humanity, with all the typicality and archetypically described in fiction and psychological literature, the experience of love will be deeply individual. The same can be said about the experience of death. But the law remains in force: practice is the criterion of truth.

M.: Do you have any questions about the reliability of the above main results? Do you agree with them?

Ph.: Speaking of post-mortem experiences, can you provide evidence of the post-mortem soul state? Has she found bliss, or is she experiencing "pangs of remorse"?

M.: Here one should distinguish between the events of the main post-mortem scenario (and its special cases) and the final state.

Post-mortem scenarios (Chapter 2) are a time of testing, adaptation to new conditions, losing many abilities, and discovering others. Naturally, losses are perceived very painfully, and can cause regret and repentance, accompanied by a "change of mind" - just as it is embedded in the Greek equivalent of the term "repentance" - "metanoia". Another circumstance is that both "ease", and duration, and, in part, the choice of one or another scenario depend on the experience and knowledge acquired during life. The destruction of the sensory space is also possible, which, of course, is accompanied by horror and torment. At the same time, the duration of scenarios according to Earth time is usually several weeks. Then, due to a different perception of time, it can be perceived individually as an eternity.

The final states are hypothetical. They are based on analogies with the "world of waves" and are interpreted as merging, repulsion, and transition into antiphase. The first can be understood as dissolution in the light (corresponds to Indian, but contradicts Abrahamic ideas), the second - as reincarnation, and the third - as a birth in the Anti-universe, but this is pure speculation yet.

Concerning the proof of the general post-mortem scenario and its particular cases, discussed in detail in Chapters 12 and 13, the following may be noted:

1. This is a mathematical extrapolation, the correctness of which is due to two circumstances: after the destruction of the soliton-wave system, its parts remain complex open systems and the laws of self-organization apply to them; if tests from the origin of life cycle events demonstrate the adequacy of abstractions, axioms, models, and scenarios, then we can expect that the adequacy will extend to post-mortem scenarios.
2. In the absence of experimental data, universal human experience (artifacts and sources) was used in the frame of verification methodology created - the verification showed their fairly good agreement by invariants.

Ph.: The emergence and evolution of life, with its inexpressible complexity compared to the inorganic world. Do you have confidence that modern theories of evolution can explain this? And how can the theory of self-organization help?

M.: Good, but difficult questions. Modern theories are developing in the right direction, but are still far from an exhaustive explanation of all known phenomena. This can be seen at least from the fact that there are several competing theories of both the origin of life and evolution and the disputes of their supporters are far from over. This situation can be compared with the attempts at the Grand Unification of all physical interactions: just as in physics it was possible to show a certain unity of electric and magnetic, electric and weak interactions, so the synthetic theory of evolution includes the data of Darwinism and molecular biology. But there is also a systemic level, in particular, ecological; in addition, research is actively developing and all the time gives new data, needs to be taken into account.

Regarding the theory of self-organization, here it was first necessary to test its possibilities by describing known regularities. Such a check has been completed, and now we can move on. First, use the S-space non-additivity. Self-organization utility in explaining phenomena, that are discussed in respective chapters. Now the "puzzle has taken shape" and I would like to write a book, Evolution, Disasters and Self-Organization, which would examine these processes as a whole.

Ph.: And catastrophe?

M.: Yes, and the catastrophe as a phenomenon can also be considered a result of self-organization, and from these positions, one can find its similarity with evolution (Kovalyov et al., 2023)

Ph.: Do you have any questions about the philosophical aspects of the problem?

M.: Certainly. First, there is the issue of evidence. Still, references to the opinion of authoritative researchers are too reminiscent of the discussions of medieval scholastics. Are there convincing experiments or testable models behind all this?

Ph.: Here again we return to what I said above. Religious and transpersonal experience is a self-verifying experience. Silouan the Athos has such a characteristic statement: "When God appears to an ascetic in light and fullness, then there can be no doubt that this is truly God the Almighty. God Himself testifies of Himself that it is He. The experimental evidence base is the individual passage of the spiritual life path according to ascetic laws.

M.: Secondly, it seems unreasonably optimistic to turn to transpersonal experience to solve the problem of the meaning of being. A simple example of such an experience - dreamless sleep - believed to restore various psycho-physiological structures, does not give any spiritual knowledge. In the same way, states of trance or near-death experiences should be assessed as a result of pharmacological influences, injuries, and diseases. Moreover, transpersonal experience is dangerous precisely in a spiritual sense - remaining within the framework of the Orthodox tradition, let's name such authors as Rev. John of the Ladder, Rev. Simeon the New Theologian, St. Ignatius (Bryanchaninov), teacher Silouan of Athos, directly pointing to the danger of such a path, and many others can be included in this list. Also, for example, in the Kyivo-Pechersk Patericon, there are many stories about fallen, mad, seriously ill monks who decided to engage in such practices without experienced mentors. The entire history of mankind shows that transpersonal experience can be adopted only by very few, and it can solve the problem of the meaning of existence loss only at the individual level, but not in general. What do you think about it?

Ph.: There are criteria for distinguishing true spiritual experience from false experience and delusion. They are described in sufficient detail in ascetic literature. And these dangers you speak of can be avoided. Transpersonal experiences are for the few, but the driving force behind history is an active creative minority, not a passive, the silent majority. Therefore, it is necessary to solve the problem of meaning with them.

M.: Thirdly, the attitude towards death as a self-sufficient initiatory event seems doubtful. For a trained person, any event can be initiatory - coming of age, obtaining a driver's license, defending a dissertation, and so on; an unprepared person will perceive them only as bureaucratic procedures that complicate life. For most Ancient Egyptians, who paid more attention to death, than any other people, the afterlife was an improved continuation of earthly life. For them, death was not an initiation. Another thing is the priests and initiates, who have the appropriate personal qualities, and potential and have undergone the necessary training. It is essential to consider the psychotype, the potential of the individual, and the features of consciousness depending on the historical era, cultural context, and age characteristics. According to the definitions given in chapters 3 and 9, all this is knowledge of the 1st and 2nd kinds, these are calibrations, that prepare initiation the result of which is the ability to intuitively perceive in the "direction" given by the calibrations. Do you agree with this clarification?

Ph.: I agree. Not every person is ready and can initiate. Metaphysical nostalgia is also not inherent in everyone. Rather, today it is more of an exception to the rule. But, as the Bible says, "Many are called, but few are chosen."

M.: And also, about the modern era. On the one hand, this is the time when various sensory experiences have developed and are perceived as an independent value, and this can be considered as the peak of the life cycle of humanity, which finally opened the channels of 5-6 levels (due to a decrease in the potentials of previous levels); and how it was in earlier eras is described in (Kovalyov et al., 2020). On the other hand, the surplus of potential used for the development of mankind has already been exhausted, as it happens in the life cycle of an individual when reaching maturity (Chapter 9). In the future, an alternative is possible: either to strive to maintain what has been achieved (a conservative tradition, to which, as far as I understand, you are inclined) or as a desire to "go with the flow" (liberal tradition, modern "mainstream"). I note that neither one nor the other traditions of the laws of self-organization are canceled - humanity will inevitably grow old and degrade, which is already observed. Likewise, both options are useless for individual salvation.

Ph.: The entropy of humans as a species is inevitable, you are right, however, one should not forget that civilization is born and develops in response to challenges. The historian A. Toynbee wrote about this at one time. Perhaps, we observe a relative calm and lack of challenges in some periods of history. Then humanity becomes smaller, morally degraded, and intellectually depleted. However, in the face of serious challenges (epidemics, wars, natural disasters), entropy slows down and even reverses. A person receives an existential "shake" and begins to develop.

Today, when the Third World War is going on, the general political and psychological tension in international relations can give rise to a social explosion. This revolution will lead to a change of elites and the general course of history "down". But this is just one of the possible scenarios for the future.

M.: And the last question for discussion: what does our book give a person? How useful is it? I'll try to offer my answer. The ancient Egyptian in the afterlife could count on his strength and knowledge, the help of relatives, the priests' rituals, and the gods' benevolence. An Orthodox Christian humbly hopes for the help of relatives and friends, the prayers of the church, the protection of angels, and the mercy of God. A Buddhist relies on personal knowledge and meditation techniques.

Reading this book leads to gaining knowledge. The facts and models given in it are gauges corresponding to the knowledge of the first kind; the changes in the settings of the sensory space that occurred after reading are calibrations corresponding to the knowledge of the second kind. As a result, conditions are created for constructing a ternary connective and obtaining intuitive knowledge. Suppose the reader undertakes an abstract study of the life and death problem. In that case, he will need to study all components: the evolution of living organisms, finding ways to extend life, etc. If, alas, the reader has died and is trying to navigate the new reality, then for him the completion of an intuitive act may be the desired realization of the general post-mortem scenario or one of its particular cases. In both cases, understanding the philosophical problem is necessary.

Ph.: Perhaps, it is possible to finish.

M: I agree. I think it is important for both mathematicians and philosophers to have a dialogue and find a common understanding of the most complex problem of life and death.

REFERENCES

Aristotle, . (1978). O sofisticheskih oproverzeniyah. [On sophistical rebuttals] In *Works in 4 volumes* (Vol. 2, pp. 535–593). Mysl.

Kovalyov, Y. (2023). Hermeneutic analysis in the framework of complex systems self-organization theory. *Svitovi vymiry osvitnikh tendencij: zbirnyk naukovykh pracj 16,* 99-110. https://imco.nau.edu.ua/

Kovalyov, Y., Mkhitaryan, N., & Nitsyn, A. (2020). *Self-organization of the Human Mind and the Transition from Paleolithic to Behavioral Modernity*. IGI Global International Publisher of Progressive Information Science and Technology Research. DOI: 10.4018/978-1-7998-1706-2

Kovalyov, Y., Shmelova, T., Sikirda, Y., & Yatsko, M. (2023). Catastrophe as an anti-system: Prevention of the flight emergency development. *2023 13th International Conference on Dependable Systems, Services and Technologies (DESSERT)*. https://doi.org/DOI: 10.1109/DESSERT61349.2023.10416548

Mkhitaryan, N. M., Badeyan, G. V., & Kovalyov, Y. N. (2004). *Erghonomycheskye aspekti slozhnikh system* [Ergonomic Aspects of Complex Systems]. Naukova Dumka.

Mkhitaryan, N. M., Kovalyov, Y. M., Malik, T. V., Safronov, V. K., & Safronova, O. O. (2021). *Dyzajn seredovyshha mista: Baghatokryterialjna optymizacija ta rozumni tekhnologhiji* [Design of the City Environment: Multi-Criteria Optimization and Smart Technologies]. Naukova dumka.

Compilation of References

Inanna. Author: Sailko. *File: Ishtar on an akkadian seal.Jpg*. (n.d.). Wikimedia.org. Retrieved August 15, 2024, from https://commons.wikimedia.org/wiki/File:Ishtar_on_an_Akkadian_seal.jpg

Ereshkigal. Photographer: BabelStone. *File: British museum queen of the night.Jpg*. (n.d.). Wikimedia.org. Retrieved August 15, 2024, from https://commons.wikimedia.org/wiki/File:British_Museum_Queen_of_the_Night.jpg

Gilgamesh. Photographer: U0045269. Wikipedia contributors. (n.d.). *File:O.1054 color.jpg*. Wikipedia, The Free Encyclopedia. https://en.wikipedia.org/wiki/File:O.1054_color.jpg

Abrahamsson, H. (2009). *The origin of death: studies in African mythology*. Cambridge University Press. (Original work published 1951)

Amin, O. S. M. FRCP(Glasg) (2016). Ashurnasirpal II performs religious rituals before the sacred tree. [Illustration]. *File: Ashurnasirpal II performs religious rituals before the sacred tree. From Nimrud, Iraq. 865-860 BCE. British Museum.jpg*. (n.d.). Wikimedia.org. Retrieved August 15, 2024, from https://commons.wikimedia.org/wiki/File:Ashurnasirpal_II_performs_religious_rituals_before_the_sacred_tree._From_Nimrud,_Iraq._865-860_BCE._British_Museum.jpg

Anatomy 3D Atlas (n / d). https://anatomy3datlas.com

Apuleius, L. (1989). *Metamorphoses (The Golden Ass)*. Harvard University Press.

Baloyi, L., & Makobe-Rabothata, M. (2014). The African conception of death: A cultural implication. *Toward Sustainable Development through Nurturing Diversity*. https://doi.org/DOI: 10.4087/FRDW2511

Bedau, M. (2014). *The Nature of Life*. Cambridge University Press.

Behrendt, K. (2018). Unmoored: Mortal harm and mortal fear. *Philosophical Papers*, 48(2), 179–209. DOI: 10.1080/05568641.2018.1462668

Berndt, R. M., & Berndt, C. H. (1988). *Catherine Helen (Author) The World of the First Australians: Aboriginal Traditional Life: Past and Present*. Aboriginal Studies Pr. (Original work published 1964)

Bird-man (a man with a bird's head). [Illustration]. Cave of Lascaux. France. (n.d.). Yandex.com. Retrieved August 14, 2024, from https://yandex.com/collections/card/5a9c60b22321f21565f700e4/

Blackmor, S. J. (2000). First person – into the unknown. *New Scientist*, 4, 55.

Blackmor, S. J. (2005). *Beyond The Body: An Investigation of Out-of-the-Body Experiences*. Academy Chicago Publishers.

Brouwer, L. E. J. (1907). *Over de grondslagen der wiskunde*. Amst.-Lpz.

Brouwer, L. E. J. (1908). *De onbetrouwbaarheid der logische principes* (Vol. 2). Tijdsehz voor Wijsbegeerte.

Buzsáki, G. (2007). The structure of consciousness. *Nature*, 446(7133), 267. DOI: 10.1038/446267a

Buzsáki, G., & Watson, B. O. (2012). Brain rhythms and neural syntax: Implications for efficient coding of cognitive content and neuropsychiatric disease. *Dialogues in Clinical Neuroscience*, 14(4), 345–367. DOI: 10.31887/DCNS.2012.14.4/gbuzsaki

Callistos Ware. (2000). *The Inner Kingdom: Collected Works*. ladimir's Seminary Press

Charlier, P., & Deo, S. (2018). The notion of soul and its implications on medical biology. *Ethics, Medicine, and Public Health*, 5, 125–127. DOI: 10.1016/j.jemep.2018.05.005

Charon. Illustration of Dante's Divina Commedia Author: Doré, R. G. [Illustration]. *File: Charon by Dore.Jpg*. (n.d.). Wikimedia.org. Retrieved August 14, 2024, from https://commons.wikimedia.org/wiki/File:Charon_by_Dore.jpg

Cholbi, M. (2016). *Immortality and the Philosophy of Death*. Rowman & Littlefield.

Cholbi, M. (2022). *Grief: A Philosophical Guide*. Princeton University Press. DOI: 10.2307/j.ctv1n1bs19

Cholbi, M. (2023). *Oxford Handbook of the Philosophy of Suicide*. Oxford University Press.

Christenson, A. J. (2007). *Popol Vuh: The Sacred Book of the Maya*. University of Oklahoma Press.

Coleman, G. (Ed.). Jinpa, T. (Ed.), Dorje, G. (Transl.), & Dalai Lama (Comment.). (2007). *The Tibetan Book of the Dead: First Complete Translation.* Penguin Classics

Coleman, G. (Ed.). Jinpa, T. (Ed.), Dorje, G. (Transl.), Dalai Lama (Comment.). (2007). *The Tibetan Book of the Dead: First Complete Translation.* Penguin Classics

Collins, A. (2014). *Göbekli Tepe: The Origin of the Gods.* Bear & Company.

Curry, A. (2020). DNA from ancient Irish tomb Realitys incest and an elite class that ruled early farmers. *Science.* Advance online publication. DOI: 10.1126/science.abd3676

Dasgupta, S. A. (2014). *History of Indian Philosophy.* (5 vol.). Cambridge University Press. (Original work published 1922)

De Spinoza, B. (2020). *Spinoza's Ethics.* Princeton University Press. (Original work published 1677)

Dedović, B. (2020). *"Inanna's Descent to the Netherworld": A centennial survey of scholarship, artifacts, and translations.* Digital Repository at the University of Maryland. https://doi.org/DOI: 10.13016/ur74-yqly

Dmitris1 (2013). Diffraction of sea waves at breakwater, Ashkelon, Israel. [Illustration]. *File: BreakWaterDiffraction Ashkelon1.jpg - Wikimedia Commons.* (2013, March 24). https://commons.wikimedia.org/wiki/File:BreakWaterDiffraction_Ashkelon1.jpg

Ebenstein, J. (2017). *Death: A Graveside Companion.* Thames & Hudson.

Egerstrom, K. (2021). *Making death not quite as bad for the one who dies.* Routledge.

Eliade, M. A (1981-1988). *History of Religious Ideas.* (3 vol.). University of Chicago Press.

Eliot, T. Four quartets. *T.S. Eliot: Four Quartets - an accurate online text.* (n.d.). Davidgorman.com. Retrieved August 15, 2024, from http://www.davidgorman.com/4quartets/

Evola, J. (2003). *Ride the Tiger: A Survival Manual for the Aristocrats of the Soul.* Inner Traditions/Bear.

Fischer, J. M. (2019). *Death, immortality, and meaning in life.* Oxford University Press.

Foster, B. R. (2001). *The Epic of Gilgamesh.* W.W. Norton & Company.

Frank, M. G. (2008). *Brain Rhythms*. Frank, M. G. (2008). Brain Rhythms. In *Springer eBooks* (pp. 482–483). DOI: 10.1007/978-3-540-29678-2_727

Frankl, V. (2006). *Man's Search for Meaning*. Beacon Press.

Fremantle, F. (2001). *Luminous Emptiness Understanding the Tibetan Book of the Dead*. Shambhala Publications.

Freud, S. (1978). *The Ego and the Id and Other Works (1923–26)*. v. XIX. In The Standard Edition of the Complete Psychological Works of Sigmund Freud. London the Hogarth Press. (Original work published 1923)

Gadamer, H.-G. (1989). *Truth and Method* (2nd ed.). Continuum Intl Pub Group.

George, A. (2014). *Epic of Gilgamesh*. Penguin Classics.

Gödel, K. (1995). Some basic theorems on the foundations of mathematics and their implications. In *Feferman, ed., 1995. Kurt Gödel Collected Works*, v. III, 304-323. Oxford University Press. (Original work published 1951)

Goelet Jr, O., Faulkner, R. O., Andrew, C. A. R., Gunther, J. D., & Wasserman, J. (2015). The Egyptian Book of the Dead: The Book of Going Forth by Day: The Complete Papyrus of Ani Featuring Integrated Text and Full-Color Images.

Goelet, O. (Transl.), Faulkner, R. (Transl.), Andrews, C. (Pref.), Gunther, J. D. (Intro.), & Wasserman, J. (Foreword). (2015). *Egyptian Book of the Dead: The Book of Going Forth by Day: The Complete Papyrus of Ani*. Chronicle Books.

Goelet, O. (Transl.), Faulkner, R. (Transl.), Andrews, C. (Pref.), Gunther, J. D. (Intro.), & Wasserman, J. (Foreword). (2015). *Egyptian Book of the Dead: The Book of Going Forth by Day: The Complete Papyrus of Ani*. Chronicle Books. Flammarion, C. (1920-1922). *La mort et son mystère*. Editeur Fantaisium.

Google Maps. Kyiv._https://www.google.com.ua/maps/search/zmieva+pechera+Kiev/@50.4574117,30.5183514,12.79z?hl=ru&entry=ttu

Guénon, R. (2004). *Perspectives on initiation*. Sophia Perennis.

Guyton, A. C., & Hall, J. E. (2005). *Textbook of Medical Physiology* [Textbook]. Saunders. (Original work published 1969)

Hagglund, M. (2019). *This Life: Why Mortality Makes Us Free*. Profile Books.

Hero mastering a lion. Relief. Palace of Sargon II at Khorsabad, 713–706 BC. Collection: Louvre Museum. Department of Oriental Antiquities, Richelieu, ground floor, room 4. Accession number: AO 19862. Photographer: Jastrow (2006). [Illustration]. *File: Hero lion Dur-Sharrukin Louvre AO19862.jpg.* (n.d.). Wikimedia.org. Retrieved August 14, 2024, from https://commons.wikimedia.org/wiki/File:Hero_lion_Dur-Sharrukin_Louvre_AO19862.jpg

Heroes of Sumerian-Akkadian myths. [Illustration]. Used files:

Hesiod. Theogony. Nagy, G.& Banks, J., (Trans.). (2020, November 2). The Center for Hellenic Studies. https://chs.harvard.edu/primary-source/hesiod-theogony-sb/

Heyting, A. (1930). Die formalen Regeln der intuitionistischen Logik. 3 parts, In Sitzungsberichte der preußischen Akademie der Wissenschaften. phys.-math. Klasse. 42-65, 57-71, 158–169.

Heyting, A. (1956). *Intuitionism. An introduction*. North-Holland Publishing Co.

Hilbert, D. (1950). *Grundlagen der Geometrie* [The Foundations of Geometry]. 2nd ed.). (Townsend, E. J., Trans.). Open Court Publishing. (Original work published 1902)

Hilbert, D., & Bernays, P. (1934). *Grundlagen der mathematic* [The Foundations of Mathematics]. Julius Springer.

Holy Bible - American Standard Version. (2019, April 1). Holy Bible - American Standard Version - ASV. https://holy-bible.online/asv.php?

Holy Bible - American Standard Version. (2019, April 1). Holy Bible - American Standard Version - ASV. https://holy-bible.online/asv.php? Landscape with Noah, Offering a Sacrifice of Gratitude. J.A. Koch, K.G. Chic (1803). Shtedel Art Institute. 2nd floor, Kunst der Moderne. ID 767 Source: The Yorck Project (2002). [Illustration]. *File: Joseph Anton Koch 006.jpg.* (n.d.). Wikimedia.org. Retrieved August 14, 2024, from https://commons.wikimedia.org/wiki/File:Joseph_Anton_Koch_006.jpg?uselang=ru

Homer. *The Iliad*. Butler, S. (Trans.). (N.d.). Mit.edu. Retrieved August 15, 2024, from http://classics.mit.edu/Homer/iliad.html

Johnston, S. A. (2006). *Inside the Neolithic Mind: Consciousness*. Cosmos, and the Realm of the.

Joshi, K. L. (Sanskrit Text, English Transl.). (2005). *112 Upanishads.* (2 vol.) Parimal Publications.

Jung, C.-G. (2002). Psychological Commentary on The Tibetan Book of the Dead. In *Jung on Death and Immortality*. Princeton University Press., DOI: 10.1515/9780691215990-004

Kamm, F. M. (2021). *Almost over: aging, dying, dead*. Oxford University Press.

Kaufman, F. (2016). *Lucretius and the Fear of Death*. Cholbi.

Kaufman, F. (2022). Death, deprivation, and a Sartrean account of horror. *Pacific Philosophical Quarterly*, 103(2), 335–349. DOI: 10.1111/papq.12353

Keightley, D. N. (2004). *The Making of the Ancestors: Late Shang Religion and Its Legacy*. In *Religion and Chinese Society* (Lagerwey, J. Edit.). (2 vol.). The Chinese UP DOI: 10.2307/j.ctv1z7kkfn.4

Kokosalakis, N. (2020). Reflections on Death. *Philosophical/Existential Context. Society*, 57(4), 402–409. DOI: 10.1007/s12115-020-00503-5 PMID: 32836561

Kovalyov, Y. (2023). Hermeneutic analysis in the framework of complex systems self-organization theory. *Svitovi vymiry osvitnikh tendencij: zbirnyk naukovykh pracj 16,* 99-110. https://imco.nau.edu.ua/

Kovalyov, Y. M., & Kalashnikova, V. V. (2023). Human life cycle modeling. *Suchasni problemy modeljuvannja.* [Modern Problems of Modeling], *25,* 110-122

Kovalyov, Y., Mkhitaryan, N., & Nitsyn, A. (2020). *Self-organization of the human mind and the transition from paleolithic to behavioral modernity*. IGI Global

Kovalyov, Y., Shmelova, T., Sikirda, Y., & Yatsko, M. (2023). Catastrophe as an anti-system: Prevention of the flight emergency development. *2023 13th International Conference on Dependable Systems, Services and Technologies (DESSERT)*. https://doi.org/DOI: 10.1109/DESSERT61349.2023.10416548

Kovalyov, Y., Mkhitaryan, N., & Nitsyn, A. (2020). *Self-organization of the human mind and the transition from paleolithic to behavioral modernity*. IGI Global International Publisher.

Kovalyov, Y., Mkhitaryan, N., & Nitsyn, A. (2020). *Self-organization of the Human Mind and the Transition from Paleolithic to Behavioral Modernity*. IGI Global.

Lamont, P. (2007). Paranormal Belief and the Avowal of Prior Scepticism. *Theory & Psychology*, 17(5), 681–696. DOI: 10.1177/0959354307081624

Lascaux cave. Free placement of drawings and use of the relief of the walls of the cave. [Illustration]. Lascaux cave, France. (n.d.). Yandex.ru. Retrieved August 15, 2024, from https://yandex.ru/collections/card/5a07c8eb0c1ed2002fed3a3e/

Leroi-Gourhan, A. (1982). *The Dawn of European Art: An Introduction to Palaeolithic Cave Painting*. Cambridge University Press Leroy- Gourhan, A. (1992). *L'art pariétal - Langage de la préhistoire*. Jérôme Million

Leroi-Gourhan, A. (1967). *Treasures of Prehistoric*. Harry N. Abrams.

Leroi-Gourhan, A. (2015). *Les religions de la préhistoire* (7th ed.). PUF.

Levi-Strauss, C. (1966). *The Savage Mind*. University of Chicago Press.

Lewis-William, D. (2002). *The Mind in the Cave*. Thames and Hudson.

Lhote, H. (1959). *The search for the Tassili frescoes: the story of the pre-historic rock paintings of the Sahara*. Dutton.

Manabu, F. (2023). Exploring the philosophical concept of my death in the context of biology: The scholarly significance of the unknown. *Continental Philosophy Review*, 56(2), 317–333. DOI: 10.1007/s11007-022-09596-7

Mannix, K. (2018). With the End in Mind: Dying, Death, and Wisdom in an Age of Denial. Little, Brown Spark.

Maslow, A. H. (1968). *Toward a psychology of being*. Van Nostrand.

May, R. (1996). *The Meaning of Anxiety*. Norton.

Meagher, D. K. (Ed.). (2017). *Handbook of thanatology: The essential body of knowledge for the study of death, dying, and bereavement* (2nd ed.). Routledge.

Mercer, S. A. B. (Ed.). (1952). *The Pyramid Texts*. (n.d.). Sacred-texts.com. Retrieved August 15, 2024, from https://www.sacred-texts.com/egy/pyt/index.htm

Mercer, S. A. B. (2020). *The Pyramid Texts*. Global Grey. (Original work published 1952)

Mishra, Y. (2019). Critical Analysis of Panchakosha Theory of Yoga Philosophy. *World Journal of Pharmaceutical Research*, 8(13), 413. DOI: 10.20959/wjpr201913-16152

Mkhitaryan, N. M., Badeyan, G. V., & Kovalyov, Y. N. (2004). *Erghonomycheskye aspekti slozhnikh system* [Ergonomic Aspects of Complex Systems]. Naukova Dumka. Aristotle. (1978). O sofisticheskih oproverzeniyah. [On sophistical rebuttals]. In Works in 4 volumes. - v.2., 535-593. Mysl.

Mkhitaryan, N. M., Kovalyov, Y. M., Malik, T. V., Safronov, V. K., & Safronova, O. O. (2021). *Dyzajn seredovyshha mista: Baghatokryterialjna optymizacija ta rozumni tekhnologhiji* [Design of the City Environment: Multi-Criteria Optimization and Smart Technologies]. Naukova dumka.

Morozov A. (2017). Moral and Religious Motives in the Works of J.R.R. Tolkien: Cultural Context. *Ukrainian cultural studies, 1*, 60-65

Morozov A. (2022). Death as existential problem: philosophical and religious aspects. *The journal of oriental studies. Institute of oriental philosophy* (IOP), *31*, 172-191

Morozov A. (2024). Ontological dimensions of the main ethical categories: freedom, conscience, equality and dignity. *The journal of oriental studies. Institute of oriental philosophy (IOP), 33*, 251-273

Morozov, A., Nikulchev, & M., Kulagin, Y. (2021). Religious experience and temptations of nihilism in spiritual life of personality. *Beytulhikme. The international journal of philosophy, 11*(2), 683-697.

Mwania, P. (2016). Interface between African's Concept of Death and Afterlife and the Biblical Tradition and Christianity. (N.d.). Tangaza.Ac.Ke. Retrieved August 14, 2024, from https://repository.tangaza.ac.ke/server/api/core/bitstreams/2ad4ce8c-5d88-496a-8986-2b23b084f309/content

Newgrange Tomb, Newgrange, Ireland. Approximately 3200 years. BC. [Illustration]. [Used Files]:

Newgrange Tomb. A stone with images of spirals at the entrance. [Rick Doble]. *Computing the Winter Solstice at Newgrange: Comparing Neolithic science to Greek or Roman science.* (n.d.). Newgrange.com. Retrieved August 15, 2024, from https://www.newgrange.com/winter-solstice-newgrange.htm

Newgrange Tomb. Modern view after restoration. (n.d.). [Newgrange. 2 February 2018]. Travellan.ru. Retrieved August 15, 2024, from https://travellan.ru/articles/nyugreyndzh/

Newgrange Tomb. Sun rays in the corridor on the winter solstice day. [Newgrange Winter Solstice. December 19, 2012]. *Irish History.* (n.d.). Blogspot.com. Retrieved August 15, 2024, from http://history-ireland.blogspot.com/2012/12/newgrange-winter-solstice.html

Newgrange Tomb. The design of the mound, corridor, and chamber. (N.d.). Rumpus.Ru. Retrieved August 15, 2024, from https://rumpus.ru/turizm/nyugrejnzh-kurgan-fej-v-irlandii/

Newgrange Tomb. The path of the sun's rays on the winter solstice day. [Rick Doble]. *Computing the Winter Solstice at Newgrange: Comparing Neolithic science to Greek or Roman science.* (n.d.). Newgrange.com. Retrieved August 15, 2024, from https://www.newgrange.com/winter-solstice-newgrange.htm

Newgrange Tomb. View of the central chamber. [Newgrange. 2 February 2018]. (n.d.). Travellan.ru. Retrieved August 15, 2024, from https://travellan.ru/articles/nyugreyndzh/

Noah's Ark. Islamic miniature. Before XVIII c. [Illustration]. *File: Noah islam 2.jpg*. (n.d.). Wikimedia.org. Retrieved August 14, 2024, from https://commons.wikimedia.org/wiki/File:Noah_islam_2.jpg?uselang=ru

Old European culture. (n.d.). Blogspot.com. Retrieved August 15, 2024, from https://oldeuropeanculture.blogspot.com/2016/12/newgrange.html

Otto, W. F. (1954). *The Homeric Gods*. Thames & Hudson. (Original work published 1929)

Otto, W. F. (1995). *Dionysos. Mythos und Kultus*. Indiana University Press. (Original work published 1933)

Otto, W. F. (2015). *Teofania*. Sexto Piso Espana S L. (Original work published 1956)

Parada, C., & Förlag, M. (1997). Map of the Underworld. Showing the descents of Odysseus and Aeneas. [Illustration]. *Map of the underworld - Greek Mythology Link*. (n.d.). Maicar.com. Retrieved August 15, 2024, from https://www.maicar.com/GML/Underworldmap.html

Paulme, D. (1967). Two Themes on the Origin of Death in West Africa. *Man*, 2(1), 48–61. DOI: 10.2307/2798653

Pindar, . (1947). *The extent odes of Pindar*. The University of Chicago.

Plato, . (1925). *Plato in Twelve Volumes*. Harvard University Press.

Pomeroy, E., Mirazón Lahr, M., Crivellaro, F., Farr, L., Reynolds, T., Hunt, C. O., & Barker, G. (2017). Newly discovered Neanderthal remains from Shanidar Cave, Iraqi Kurdistan, and their attribution to Shanidar 5. *Journal of Human Evolution*, 111, 102–118. DOI: 10.1016/j.jhevol.2017.07.001 PMID: 28874265

Puetz, S. J. (2022). The infinitely fractal universe paradigm and consupponibility. *Chaos, Solitons, and Fractals*, 158(112065), 112065. DOI: 10.1016/j.chaos.2022.112065

Radhakrishnan, S. (2009). *Indian Philosophy*. (2 vol.). Oxford University Press. (Original work published 1923)

Radhakrishnan, S. (2009). *Indian Philosophy* (Vols. 1–2). Oxford University Press. (Original work published 1923)

Ravi Kumar Patil, H. S., Makari, H. K., Gurumurthy, H., & Sowmya, S. V. (2009). *A Textbook of Human Physiology*. I K International Publishing House.

Riera, J. J. *Semiotic Theory*. Press Books.

Sakharov, S. (1988). *We Shall See Him as He Is*. Stravropegic Monastery of St. John the Baptist.

Satpathy, B.Dr. Biswajit Satpathy. (2018). Pancha Kosha Theory of Personality. *International Journal of Indian Psychology*, 6(2). Advance online publication. DOI: 10.25215/0602.105

Schmidt, K. (2020). *Sie bauten die ersten Tempel: Das rätselhafte Heiligtum am Göbekli Tepe*. C.H.Beck.

Schmidt, K. (2020). *Sie bauten die ersten Tempel: Das rätselhafte Heiligtum am Göbekli Tepe*. C.H. Beck.

Schmidt, R. F., & Thews, G. (Eds.). (1983). *Human Physiology*. Springer-Verlag.

Schwarz, B., & Krawczyk, C. M. (2020). Coherent diffraction imaging for enhanced fault and fracture network characterization. *Solid Earth*, 11, 1891–1907. DOI: 10.5194/se-11-1891-2020

Sethumadhavan, T. N. (2011). Aitareya Upanishad Transliterated Sanskrit. Brief Explanation. (n.d.). *Transliterated Sanskrit text free translation*. Esamskriti.com. Retrieved August 15, 2024, from https://esamskriti.com/essays/Aitareya-Upanishad.pdf

Shmelova, T., Sikirda, Y., Rizun, N., Salem, A.-B. M., & Kovalyov, Y. (2018). *Socio-Technical Decision Support in Air Navigation Systems: Emerging Research and Opportunities*. IGI Global. DOI: 10.4018/978-1-5225-3108-1

Silouan the Athonite. (1991). *The Life and Teachings*. Palomnik. Hegel, G. (1977). *Phenomenology of spirit*. Oxford university press Heidegger, M. (2006). *Sein und Zeit*. Niemeyer, Tübingen. (Original work published 1927) Nagy. Gregory. (2018). *Homeric hymn to Demeter*. University of Houston Press. Kovalyov, Y., Mkhitaryan, N., & Nitsyn, A. (2020). *Self-organization of the Human Mind and the Transition from Paleolithic to Behavioral Modernity*. IGI Global International Publisher of Progressive Information Science and Technology Research.

Sophocles, . (1906). Tragedies and fragments. *Health*.

St. John Chrysostom. Archbishop of Constantinople. *The Paschal Sermon*. (n.d.). Oca.org. Retrieved August 15, 2024, from https://www.oca.org/fs/sermons/the-paschal-sermon

St. John of Damascus. (2008). *The funeral hymns of St. John of Damascus*. (n.d.). Blogspot.com. Retrieved August 15, 2024, from_http://full-of-grace-and-truth.blogspot.com/2008/12/funeral-troparia-of-st-john-of-damascus.html

Swami Nikhilananda. (Transl.). (1986-1994). Upanishads. (4 vol.). Ramakrishna Vivekanada Center

Tann. (2019). A detail from the floor of a coffin of Gua, physician of Djehutyhotep, a nomarch of Deir el-Bersha, Egypt, during the Middle Kingdom. [Illustration]. Source: Remains of 4000-year-old Egyptian Guide. Archaeonewsnet Date: 2019 Author: Tann. https://archaeonewsnet.com/2019/12/remains-of-4000-year-old-egyptian-guide.html

Tarski, A. (1969). Truth and Proof. *Scientific American* 220: 63–77 (Original work published 1936) The Human Body. (n/d). https://www.healthline.com/human-body-maps

Taylor, E. B. (2010). *Primitive Culture. Researches into the Development of Mythology, Philosophy, Religion, Art, and Custom*. Cambridge University Press. (Original work published 1871), DOI: 10.1017/CBO9780511705960

Taylor, J. S. (2014). *The Metaphysics and Ethics of Death*. Oxford University Press.

The Deluge. R. G. Dore. [Illustration]. *File: World destroyed by water.Png*. (n.d.). Wikimedia.org. Retrieved August 14, 2024, from https://commons.wikimedia.org/wiki/File:World_Destroyed_by_Water.png

The judgment of the dead in the presence of Osiris. [Illustration]. Papyrus of Hunefer. Ancient Egypt. Collection. British Museum. Accession number EA 9901. Author: Unknown author. Wikipedia contributors. (n.d.). *File: The judgement of the dead in the presence of Osiris.jpg*. Wikipedia, The Free Encyclopedia. https://en.wikipedia.org/wiki/File:The_judgement_of_the_dead_in_the_presence_of_Osiris.jpg

The Ordeal of St. Theodora. (N.d). Archaeonewsnet.com. Retrieved August 15, 2024, from https://katolyki-krasnodara.ru/en/mytarstva-prepodobnoi-feodory-mytarstva-blazhennoi-feodory-chto.html

Theudbald. (2006). River Acheron. [Illustration]. Date: 2006. Author: Theudbald. (N.d). Wikimedia.org. Retrieved August 15, 2024, from https://commons.wikimedia.org/wiki/File:Acheron.JPG?uselang=en

Thurman, R. (2011). *The Tibetan Book of the Dead: Liberation Through Understanding in the Between*. Bantam.

Tillich, P. (2000). *The courage to be*. Yale University Press.

Tononi, G., & Edelman, G. M. (1998). Consciousness and complexity. *Science*. *4*(282) (5395),1846-51. DOI: 10.1126/science.282.5395.1846

Van Dijk, T. A. (2006). Discourse, context and cognition. *Discourse Studies*, 8(1), 159–177. Advance online publication. DOI: 10.1177/1461445606059565

Veerendra. (2022, November 18). *Analysing interference of waves*. [Illustration]. A Plus Topper; Aplus Topper. https://www.aplustopper.com/analysing-interference-waves/

Virgil. *The Aeneid*. Dryden, J. (Trans.). (N.d.). Mit.edu. Retrieved August 15, 2024, from http://classics.mit.edu/Virgil/aeneid.html

Wallis Budge, E. A. (1967). *The Book of the Dead. The Papyrus of Ani*. Dover Publications. (Original work published 1895)

Watkins, T. (2005). *From Foragers to Complex Societies in Southwest Asia*. Kapitel 6. In *The Human Past: World Prehistory & the Development of Human Societies*. Thames & Hudson.

Wikipedia contributors. (n.d.). *File: Aboriginal rock art on the Barnett River, Mount Elizabeth Station.jpg*. [Illustration]. Author: Graeme Churchard. Wikipedia, The Free Encyclopedia. Retrieved September, 2, 2024. https://en.wikipedia.org/wiki/File:Aboriginal_rock_art_on_the_Barnett_River,_Mount_Elizabeth_Station.jpg

Wittkowski, J., Doka, K. J., Neimeyer, R. A., & Vallerga, M. (2015). Publication trends in thanatology: An analysis of leading journals. *Death Studies*, 39(8), 453–462. DOI: 10.1080/07481187.2014.1000054

Wolff, B. M. (2015). A review of "body farm" research facilities across America with a focus on policy and the impacts when dealing with decompositional changes in human remains. [Thesis]. The University of Texas at Arlington. http://hdl.handle.net/10106/25510

Wright, M. R. (1985). *The presocratics: The main fragments in Greek with introduction, commentary and appendix containing text and translation of Aristotle on Presocratics*. Bristol Classical Press.

Yourgrau, P. (2019). *Death and Nonexistence*. Oxford University Press. DOI: 10.1093/oso/9780190247478.001.0001

Zoja, L. (2000). *Drugs, addiction, and initiation: modern search for ritual*. Daimon Verlag.

Related References

To continue our tradition of advancing information science and technology research, we have compiled a list of recommended IGI Global readings. These references will provide additional information and guidance to further enrich your knowledge and assist you with your own research and future publications.

Abu Seman, S. A., & Ramayah, T. (2017). Are We Ready to App?: A Study on mHealth Apps, Its Future, and Trends in Malaysia Context. In Pelet, J. (Ed.), *Mobile Platforms, Design, and Apps for Social Commerce* (pp. 69–83). Hershey, PA: IGI Global. DOI: 10.4018/978-1-5225-2469-4.ch005

Adeleke, I. T., & Abdul, Q. B. (2020). Opinions on Cyber Security, Electronic Health Records, and Medical Confidentiality: Emerging Issues on Internet of Medical Things From Nigeria. In Pankajavalli, P., & Karthick, G. (Eds.), *Incorporating the Internet of Things in Healthcare Applications and Wearable Devices* (pp. 199–211). IGI Global. https://doi.org/10.4018/978-1-7998-1090-2.ch012

Aggarwal, S., & Azad, V. (2017). A Hybrid System Based on FMM and MLP to Diagnose Heart Disease. In Bhattacharyya, S., De, S., Pan, I., & Dutta, P. (Eds.), *Intelligent Multidimensional Data Clustering and Analysis* (pp. 293–325). Hershey, PA: IGI Global. DOI: 10.4018/978-1-5225-1776-4.ch011

Al-Busaidi, S. S. (2018). Interdisciplinary Relationships Between Medicine and Social Sciences. In Al-Suqri, M., Al-Kindi, A., AlKindi, S., & Saleem, N. (Eds.), *Promoting Interdisciplinarity in Knowledge Generation and Problem Solving* (pp. 124–137). Hershey, PA: IGI Global. DOI: 10.4018/978-1-5225-3878-3.ch009

Al Kareh, T., & Thoumy, M. (2018). The Impact of Health Information Digitization on the Physiotherapist-Patient Relationship: A Pilot Study of the Lebanese Community. *International Journal of Healthcare Information Systems and Informatics*, 13(2), 29–53. DOI: 10.4018/IJHISI.2018040103

Alenzuela, R. (2017). Research, Leadership, and Resource-Sharing Initiatives: The Role of Local Library Consortia in Access to Medical Information. In Ram, S. (Ed.), *Library and Information Services for Bioinformatics Education and Research* (pp. 199–211). Hershey, PA: IGI Global. DOI: 10.4018/978-1-5225-1871-6.ch012

Ali, S., Samie, A. U., Ali, A., Bhat, A. H., Mir, T., & Prasad, B. V. (2020). Mental Health: A Global Issue Affecting the Pattern of Life in Kashmir. In Taukeni, S. (Ed.), *Biopsychosocial Perspectives and Practices for Addressing Communicable and Non-Communicable Diseases* (pp. 18–29). IGI Global. https://doi.org/10.4018/978-1-7998-2139-7.ch002

Almpani, S., Stefaneas, P., Boley, H., Mitsikas, T., & Frangos, P. (2019). A Rule-Based Model for Compliance of Medical Devices Applied to the European Market. *International Journal of Extreme Automation and Connectivity in Healthcare*, 1(2), 56–78. https://doi.org/10.4018/IJEACH.2019070104

Alnakhi, W. K. (2021). Medical Travel and Public Health: Definitions, Frameworks, and Future Research. In Singh, M., & Kumaran, S. (Eds.), *Growth of the Medical Tourism Industry and Its Impact on Society: Emerging Research and Opportunities* (pp. 74–94). IGI Global. https://doi.org/10.4018/978-1-7998-3427-4.ch004

Ameri, H., Alizadeh, S., & Noughabi, E. A. (2017). Application of Data Mining Techniques in Clinical Decision Making: A Literature Review and Classification. In Noughabi, E., Raahemi, B., Albadvi, A., & Far, B. (Eds.), *Handbook of Research on Data Science for Effective Healthcare Practice and Administration* (pp. 257–295). Hershey, PA: IGI Global. DOI: 10.4018/978-1-5225-2515-8.ch012

Anand, S. (2017). Medical Image Enhancement Using Edge Information-Based Methods. In Singh, B. (Ed.), *Computational Tools and Techniques for Biomedical Signal Processing* (pp. 123–148). Hershey, PA: IGI Global. DOI: 10.4018/978-1-5225-0660-7.ch006

Anton, J. L., Soriano, J. V., Martinez, M. I., & Garcia, F. B. (2018). Comprehensive E-Learning Appraisal System. In M. Khosrow-Pour, D.B.A. (Ed.), *Encyclopedia of Information Science and Technology, Fourth Edition* (pp. 5787-5799). Hershey, PA: IGI Global. https://doi.org/DOI: 10.4018/978-1-5225-2255-3.ch503

Aslan, F. (2021). Could There Be an Alternative Method of Media Literacy in Promoting Health in Children and Adolescents? Media Literacy and Health Promotion. In G. Sarı (Eds.), *Handbook of Research on Representing Health and Medicine in Modern Media* (pp. 191-199). IGI Global. https://doi.org/10.4018/978-1-7998-6825-5.ch013

Atzori, B., Hoffman, H. G., Vagnoli, L., Messeri, A., & Grotto, R. L. (2018). Virtual Reality as Distraction Technique for Pain Management in Children and Adolescents. In M. Khosrow-Pour, D.B.A. (Ed.), *Encyclopedia of Information Science and Technology, Fourth Edition* (pp. 5955-5965). Hershey, PA: IGI Global. DOI: 10.4018/978-1-5225-2255-3.ch518

Audibert, M., Mathonnat, J., Pélissier, A., & Huang, X. X. (2018). The Impact of the New Rural Cooperative Medical Scheme on Township Hospitals' Utilization and Income Structure in Weifang Prefecture, China. In I. Management Association (Ed.), *Health Economics and Healthcare Reform: Breakthroughs in Research and Practice* (pp. 109-121). Hershey, PA: IGI Global. https://doi.org/DOI: 10.4018/978-1-5225-3168-5.ch007

Aulia, A., & Pratiti, B. (2022). Cyberchondria and Medical Student Syndrome: An Anxious Path to Be an Anxiety Healer? In Aker, H., & Aiken, M. (Eds.), *Handbook of Research on Cyberchondria, Health Literacy, and the Role of Media in Society's Perception of Medical Information* (pp. 98–113). IGI Global. https://doi.org/10.4018/978-1-7998-8630-3.ch006

Babu, A., & Ayyappan, S. (2017). A Methodological Evaluation of Crypto-Watermarking System for Medical Images. In Bhatt, C., & Peddoju, S. (Eds.), *Cloud Computing Systems and Applications in Healthcare* (pp. 189–217). Hershey, PA: IGI Global. DOI: 10.4018/978-1-5225-1002-4.ch010

Bakke, A. (2017). Ethos in E-Health: From Informational to Interactive Websites. In Folk, M., & Apostel, S. (Eds.), *Establishing and Evaluating Digital Ethos and Online Credibility* (pp. 85–103). Hershey, PA: IGI Global. DOI: 10.4018/978-1-5225-1072-7.ch005

Bali, S. (2018). Enhancing the Reach of Health Care Through Telemedicine: Status and New Possibilities in Developing Countries. In Pandey, U., & Indrakanti, V. (Eds.), *Open and Distance Learning Initiatives for Sustainable Development* (pp. 339–354). Hershey, PA: IGI Global. DOI: 10.4018/978-1-5225-2621-6.ch019

Balls-Berry, J. E., & Albold, C. (2019). Asynchronous Education for Graduate Medical Trainees to Reduce Health Disparities and Address Social Determinants of Health: Online Education for Graduate Medical Trainees. In Demiroz, E., & Waldman, S. (Eds.), *Optimizing Medical Education With Instructional Technology* (pp. 1–20). IGI Global. https://doi.org/10.4018/978-1-5225-6289-4.ch001

Barrett, T. E. (2018). Essentials for Education and Training for Tomorrow's Physicians. In Smith, C. (Ed.), *Exploring the Pressures of Medical Education From a Mental Health and Wellness Perspective* (pp. 230–252). Hershey, PA: IGI Global. DOI: 10.4018/978-1-5225-2811-1.ch010

Bazan, V., Jax, M. D., & Zwischenberger, J. B. (2020). Alice in Simulation-Land: Surgical Simulation in Medical Education. In Gotian, R., Kang, Y., & Safdieh, J. (Eds.), *Handbook of Research on the Efficacy of Training Programs and Systems in Medical Education* (pp. 449–458). IGI Global. https://doi.org/10.4018/978-1-7998-1468-9.ch023

Beaudoin, S. (2019). The Hidden Face of Medical Intervention. In C. Albuquerque (Eds.), *Emerging Trends and Innovations in Privacy and Health Information Management* (pp. 188-211). IGI Global. https://doi.org/10.4018/978-1-5225-8470-4.ch008

Bernardes, O., Amorim, V., & Usman, B. (2021). The Good, the Bad, and the Ugly: Don't Blame COVID-19 for Health System Inefficiencies. In J. Santos, & I. Pereira (Ed.), *Management and Marketing for Improved Competitiveness and Performance in the Healthcare Sector* (pp. 115-139). IGI Global. https://doi.org/10.4018/978-1-7998-7263-4.ch006

Berrahal, S., & Boudriga, N. (2017). The Risks of Wearable Technologies to Individuals and Organizations. In Marrington, A., Kerr, D., & Gammack, J. (Eds.), *Managing Security Issues and the Hidden Dangers of Wearable Technologies* (pp. 18–46). Hershey, PA: IGI Global. DOI: 10.4018/978-1-5225-1016-1.ch002

Bhargava, P. (2018). Blended Learning: An Effective Application to Clinical Teaching. In Tang, S., & Lim, C. (Eds.), *Preparing the Next Generation of Teachers for 21st Century Education* (pp. 302–320). Hershey, PA: IGI Global. DOI: 10.4018/978-1-5225-4080-9.ch018

Bird, J. L. (2017). Writing Healing Narratives. In Bryan, V., & Bird, J. (Eds.), *Healthcare Community Synergism between Patients, Practitioners, and Researchers* (pp. 1–28). Hershey, PA: IGI Global. DOI: 10.4018/978-1-5225-0640-9.ch001

Bougoulias, K., Kouris, I., Prasinos, M., Giokas, K., & Koutsouris, D. (2017). Ob/Gyn EMR Software: A Solution for Obstetricians and Gynecologists. In Moumtzoglou, A. (Ed.), *Design, Development, and Integration of Reliable Electronic Healthcare Platforms* (pp. 101–111). Hershey, PA: IGI Global. DOI: 10.4018/978-1-5225-1724-5.ch006

Bouslimi, R., Ayadi, M. G., & Akaichi, J. (2018). Medical Image Retrieval in Healthcare Social Networks. *International Journal of Healthcare Information Systems and Informatics*, 13(2), 13–28. DOI: 10.4018/IJHISI.2018040102

Bouzaabia, O., Bouzaabia, R., & Mejri, K. (2017). Role of Internet in the Development of Medical Tourism Service in Tunisia. In Capatina, A., & Rancati, E. (Eds.), *Key Challenges and Opportunities in Web Entrepreneurship* (pp. 211–241). Hershey, PA: IGI Global. DOI: 10.4018/978-1-5225-2466-3.ch009

Brown-Jackson, K. L. (2018). Telemedicine and Telehealth: Academics Engaging the Community in a Call to Action. In Burton, S. (Ed.), *Engaged Scholarship and Civic Responsibility in Higher Education* (pp. 166–193). Hershey, PA: IGI Global. DOI: 10.4018/978-1-5225-3649-9.ch008

Bwalya, K. J. (2017). Next Wave of Tele-Medicine: Virtual Presence of Medical Personnel. In Moahi, K., Bwalya, K., & Sebina, P. (Eds.), *Health Information Systems and the Advancement of Medical Practice in Developing Countries* (pp. 168–180). Hershey, PA: IGI Global. DOI: 10.4018/978-1-5225-2262-1.ch010

Celik, G. (2018). Determining Headache Diseases With Genetic Algorithm. In Kose, U., Guraksin, G., & Deperlioglu, O. (Eds.), *Nature-Inspired Intelligent Techniques for Solving Biomedical Engineering Problems* (pp. 249–262). Hershey, PA: IGI Global. DOI: 10.4018/978-1-5225-4769-3.ch012

Chakraborty, S., Chatterjee, S., Ashour, A. S., Mali, K., & Dey, N. (2018). Intelligent Computing in Medical Imaging: A Study. In Dey, N. (Ed.), *Advancements in Applied Metaheuristic Computing* (pp. 143–163). Hershey, PA: IGI Global. DOI: 10.4018/978-1-5225-4151-6.ch006

Chandra, L., Downey, C. L., Svavarsdottir, H. S., Skinner, H., & Young, A. L. (2020). Gender Equity in Medical Leadership. In Bellini, M., & Papalois, V. (Eds.), *Gender Equity in the Medical Profession* (pp. 94–109). IGI Global. https://doi.org/10.4018/978-1-5225-9599-1.ch007

Chandran, S. (2019). Medical Data Storage and Compression. In Swarnambiga, A. (Ed.), *Medical Image Processing for Improved Clinical Diagnosis* (pp. 140–154). IGI Global. https://doi.org/10.4018/978-1-5225-5876-7.ch007

Chavez, A., & Kovarik, C. (2017). Open Source Technology for Medical Practice in Developing Countries. In Moahi, K., Bwalya, K., & Sebina, P. (Eds.), *Health Information Systems and the Advancement of Medical Practice in Developing Countries* (pp. 33–59). Hershey, PA: IGI Global. DOI: 10.4018/978-1-5225-2262-1.ch003

Chen, S., Traba, C., Lamba, S., & Soto-Greene, M. (2020). Professional and Career Development of Medical Students. In Gotian, R., Kang, Y., & Safdieh, J. (Eds.), *Handbook of Research on the Efficacy of Training Programs and Systems in Medical Education* (pp. 305–326). IGI Global. https://doi.org/10.4018/978-1-7998-1468-9.ch016

Chien, G. C., Haroutunian, A., England, B., & Candido, K. D. (2019). Optimizing Medical Education With Instructional Technology: Technology to Optimize Teaching Human Anatomy. In Demiroz, E., & Waldman, S. (Eds.), *Optimizing Medical Education With Instructional Technology* (pp. 71–78). IGI Global. https://doi.org/10.4018/978-1-5225-6289-4.ch005

Choudhari, R. H. (2021). Multidimensional Impact of Climate Change on Human Reproduction and Fertility: A Medical Perspective on Changing Dynamics. In K. Wani, & N. Naha (Eds.), *Climate Change and Its Impact on Fertility* (pp. 278-315). IGI Global. https://doi.org/10.4018/978-1-7998-4480-8.ch014

Ciufudean, C. (2018). Innovative Formalism for Biological Data Analysis. In M. Khosrow-Pour, D.B.A. (Ed.), *Encyclopedia of Information Science and Technology, Fourth Edition* (pp. 1814-1824). Hershey, PA: IGI Global. DOI: 10.4018/978-1-5225-2255-3.ch158

Colaguori, R., & Danesi, M. (2017). Medical Semiotics: A Revisitation and an Exhortation. *International Journal of Semiotics and Visual Rhetoric*, 1(1), 11–18. DOI: 10.4018/IJSVR.2017010102

Cole, A. W., & Salek, T. A. (2017). Adopting a Parasocial Connection to Overcome Professional Kakoethos in Online Health Information. In Folk, M., & Apostel, S. (Eds.), *Establishing and Evaluating Digital Ethos and Online Credibility* (pp. 104–120). Hershey, PA: IGI Global. DOI: 10.4018/978-1-5225-1072-7.ch006

Deperlioglu, O. (2018). Intelligent Techniques Inspired by Nature and Used in Biomedical Engineering. In Kose, U., Guraksin, G., & Deperlioglu, O. (Eds.), *Nature-Inspired Intelligent Techniques for Solving Biomedical Engineering Problems* (pp. 51–77). Hershey, PA: IGI Global. DOI: 10.4018/978-1-5225-4769-3.ch003

Desai, K. P., Shah, M. A., Lapasia, M. C., Patil, S. A., & Pathak, S. P. (2022). Applied Intelligence for Mental Health Detection: ManoVaidya – A Mental Health Therapist. In Thakare, A., Wagh, S., Bhende, M., Anter, A., & Gao, X. (Eds.), *Handbook of Research on Applied Intelligence for Health and Clinical Informatics* (pp. 80–91). IGI Global. https://doi.org/10.4018/978-1-7998-7709-7.ch005

Dey, N., & Ashour, A. S. (2018). Meta-Heuristic Algorithms in Medical Image Segmentation: A Review. In Dey, N. (Ed.), *Advancements in Applied Metaheuristic Computing* (pp. 185–203). Hershey, PA: IGI Global. DOI: 10.4018/978-1-5225-4151-6.ch008

Dey, N., Ashour, A. S., & Althoupety, A. S. (2017). Thermal Imaging in Medical Science. In Santhi, V. (Ed.), *Recent Advances in Applied Thermal Imaging for Industrial Applications* (pp. 87–117). Hershey, PA: IGI Global. DOI: 10.4018/978-1-5225-2423-6.ch004

Di Virgilio, F., Camillo, A. A., & Camillo, I. C. (2017). The Impact of Social Network on Italian Users' Behavioural Intention for the Choice of a Medical Tourist Destination. *International Journal of Tourism and Hospitality Management in the Digital Age*, 1(1), 36–49. DOI: 10.4018/IJTHMDA.2017010103

Dobani, F., Pennington, M. L., Coe, E., Morrison, P., & Gulliver, S. B. (2020). Bridging the Gaps: Toward Effective Collaboration Between Peer Supporters and Behavioral Health Professionals. In Bowers, C., Beidel, D., & Marks, M. (Eds.), *Mental Health Intervention and Treatment of First Responders and Emergency Workers* (pp. 190–204). IGI Global. https://doi.org/10.4018/978-1-5225-9803-9.ch011

Dogra, A. K., & Dogra, P. (2017). The Medical Tourism Industry in the BRIC Nations: An Indian Analysis. In Dhiman, M. (Ed.), *Opportunities and Challenges for Tourism and Hospitality in the BRIC Nations* (pp. 320–336). Hershey, PA: IGI Global. DOI: 10.4018/978-1-5225-0708-6.ch020

Doublestein, B. A., Lee, W. T., & Pfohl, R. M. (2020). Overview of Professionalism Competence: Bringing Balance to the Medical Education Continuum. In Selladurai, R., Hobson, C., Selladurai, R., & Greer, A. (Eds.), *Evaluating Challenges and Opportunities for Healthcare Reform* (pp. 215–231). IGI Global. https://doi.org/10.4018/978-1-7998-2949-2.ch010

Drowos, J. L., & Wood, S. K. (2017). Preparing Future Physicians to Adapt to the Changing Health Care System: Promoting Humanism through Curricular Design. In Bryan, V., & Bird, J. (Eds.), *Healthcare Community Synergism between Patients, Practitioners, and Researchers* (pp. 106–125). Hershey, PA: IGI Global. DOI: 10.4018/978-1-5225-0640-9.ch006

Dulam, K. (2017). Medical Patents and Impact on Availability and Affordability of Essential Medicines in India. In Aggarwal, R., & Kaur, R. (Eds.), *Patent Law and Intellectual Property in the Medical Field* (pp. 41–57). Hershey, PA: IGI Global. DOI: 10.4018/978-1-5225-2414-4.ch003

Dutta, P. (2017). Decision Making in Medical Diagnosis via Distance Measures on Interval Valued Fuzzy Sets. *International Journal of System Dynamics Applications*, 6(4), 63–83. DOI: 10.4018/IJSDA.2017100104

Dutta, P. (2021). Medical Pre-Screening of Common Diseases: An Interval-Valued Fuzzy Set Approach. In Azar, A., & Kamal, N. (Eds.), *Handbook of Research on Modeling, Analysis, and Control of Complex Systems* (pp. 267–292). IGI Global. https://doi.org/10.4018/978-1-7998-5788-4.ch011

Dwivedi, M. K., Pandey, S. K., & Singh, P. K. (2021). Public Health Surveillance System: Infectious Diseases. In Yadav, D., Bansal, A., Bhatia, M., Hooda, M., & Morato, J. (Eds.), *Diagnostic Applications of Health Intelligence and Surveillance Systems* (pp. 201–220). IGI Global. https://doi.org/10.4018/978-1-7998-6527-8.ch010

El Guemhioui, K., & Demurjian, S. A. (2017). Semantic Reconciliation of Electronic Health Records Using Semantic Web Technologies. *International Journal of Information Technology and Web Engineering*, 12(2), 26–48. DOI: 10.4018/IJITWE.2017040102

Etim, A., Etim, D. N., & Scott, J. (2020). Mobile Health and Telemedicine: Awareness, Adoption and Importance of Health Study. *International Journal of Healthcare Information Systems and Informatics*, 15(1), 81–96. https://doi.org/10.4018/IJHISI.2020010105

Fino, E., & Hanna-Khalil, B. (2020). Psychometric Post-Examination Analysis in Medical Education Training Programs. In Gotian, R., Kang, Y., & Safdieh, J. (Eds.), *Handbook of Research on the Efficacy of Training Programs and Systems in Medical Education* (pp. 221–242). IGI Global. https://doi.org/10.4018/978-1-7998-1468-9.ch012

Fishback, J. L., & Klein, R. M. (2019). Collaborative Learning for Histopathology Education. In Demiroz, E., & Waldman, S. (Eds.), *Optimizing Medical Education With Instructional Technology* (pp. 177–193). IGI Global. https://doi.org/10.4018/978-1-5225-6289-4.ch010

Fisher, J. (2018). Sociological Perspectives on Improving Medical Diagnosis Emphasizing CAD. In M. Khosrow-Pour, D.B.A. (Ed.), *Encyclopedia of Information Science and Technology, Fourth Edition* (pp. 1017-1024). Hershey, PA: IGI Global. DOI: 10.4018/978-1-5225-2255-3.ch088

Franco Santos, D., Branco Silva, A. R., Novo, M. D., & Vaz de Almeida, C. (2022). Medical Burnout: Is Mindfulness an Effective Coping Strategy? In Gupta, S. (Ed.), *Handbook of Research on Clinical Applications of Meditation and Mindfulness-Based Interventions in Mental Health* (pp. 184–194). IGI Global. https://doi.org/10.4018/978-1-7998-8682-2.ch012

Frank, E. M. (2018). Healthcare Education: Integrating Simulation Technologies. In Bryan, V., Musgrove, A., & Powers, J. (Eds.), *Handbook of Research on Human Development in the Digital Age* (pp. 163–182). Hershey, PA: IGI Global. DOI: 10.4018/978-1-5225-2838-8.ch008

Gambhir, P. (2021). Enhancing Mental Health: A Role for Technology. In Hooke, A. (Ed.), *Technological Breakthroughs and Future Business Opportunities in Education, Health, and Outer Space* (pp. 264–269). IGI Global. https://doi.org/10.4018/978-1-7998-6772-2.ch017

Garcia-Santa, N., San Miguel, B., & Ugai, T. (2019). Converging Semantic Knowledge and Deep Learning for Medical Coding. *International Journal of Privacy and Health Information Management*, 7(2), 33–52. https://doi.org/10.4018/IJPHIM.2019070103

Garner, G. (2018). Foundations for Yoga Practice in Rehabilitation. In Telles, S., & Singh, N. (Eds.), *Research-Based Perspectives on the Psychophysiology of Yoga* (pp. 263–307). Hershey, PA: IGI Global. DOI: 10.4018/978-1-5225-2788-6.ch015

Garrote Andújar, S. A., Araújo Vila, N., Brea, J. A., & de Araújo, A. F. (2021). Medical Tourism: Analysis and Expectations Worldwide. In Borges, A., & Rodrigues, P. (Eds.), *New Techniques for Brand Management in the Healthcare Sector* (pp. 84–102). IGI Global. https://doi.org/10.4018/978-1-7998-3034-4.ch006

Gavurová, B., Kováč, V., & Šoltés, M. (2018). Medical Equipment and Economic Determinants of Its Structure and Regulation in the Slovak Republic. In M. Khosrow-Pour, D.B.A. (Ed.), *Encyclopedia of Information Science and Technology, Fourth Edition* (pp. 5841-5852). Hershey, PA: IGI Global. DOI: 10.4018/978-1-5225-2255-3.ch508

Ge, X., Wang, Q., Huang, K., Law, V., & Thomas, D. C. (2017). Designing Simulated Learning Environments and Facilitating Authentic Learning Experiences in Medical Education. In Stefaniak, J. (Ed.), *Advancing Medical Education Through Strategic Instructional Design* (pp. 77–100). Hershey, PA: IGI Global. DOI: 10.4018/978-1-5225-2098-6.ch004

Gewald, H., & Gewald, C. (2018). Inhibitors of Physicians' Use of Mandatory Hospital Information Systems (HIS). *International Journal of Healthcare Information Systems and Informatics*, 13(1), 29–44. DOI: 10.4018/IJHISI.2018010103

Ghosh, D., & Dinda, S. (2017). Health Infrastructure and Economic Development in India. In Das, R. (Ed.), *Social, Health, and Environmental Infrastructures for Economic Growth* (pp. 99–119). Hershey, PA: IGI Global. DOI: 10.4018/978-1-5225-2364-2.ch006

Ghosh, K., & Sen, K. C. (2017). The Potential of Crowdsourcing in the Health Care Industry. In Wickramasinghe, N. (Ed.), *Handbook of Research on Healthcare Administration and Management* (pp. 418–427). Hershey, PA: IGI Global. DOI: 10.4018/978-1-5225-0920-2.ch024

Gonzalez-Urquijo, M., Macias-Rodriguez, Y., & Davila-Rivas, J. A. (2022). The Role of Telemedicine and Globalization in Medical Education. In Lopez, M. (Ed.), *Advancing Health Education With Telemedicine* (pp. 288–295). IGI Global. https://doi.org/10.4018/978-1-7998-8783-6.ch015

Gopalan, V., Chan, E., & Ho, D. T. (2018). Deliberate Self-Harm and Suicide Ideology in Medical Students. In Smith, C. (Ed.), *Exploring the Pressures of Medical Education From a Mental Health and Wellness Perspective* (pp. 122–143). Hershey, PA: IGI Global. DOI: 10.4018/978-1-5225-2811-1.ch005

Goswami, N. (2021). Effects of Spaceflight, Aging, and Bedrest on Falls: Aging Meets Spaceflight! In Eklund, P. (Ed.), *Integrated Care and Fall Prevention in Active and Healthy Aging* (pp. 91–106). IGI Global. https://doi.org/10.4018/978-1-7998-4411-2.ch005

Goswami, S., Dey, U., Roy, P., Ashour, A., & Dey, N. (2017). Medical Video Processing: Concept and Applications. In Dey, N., Ashour, A., & Patra, P. (Eds.), *Feature Detectors and Motion Detection in Video Processing* (pp. 1–17). Hershey, PA: IGI Global. DOI: 10.4018/978-1-5225-1025-3.ch001

Goswami, S., Mahanta, K., Goswami, S., Jigdung, T., & Devi, T. P. (2018). Ageing and Cancer: The Epigenetic Basis, Alternative Treatment, and Care. In Prasad, B., & Akbar, S. (Eds.), *Handbook of Research on Geriatric Health, Treatment, and Care* (pp. 206–235). Hershey, PA: IGI Global. DOI: 10.4018/978-1-5225-3480-8.ch012

Gouva, M. I. (2017). The Psychological Impact of Medical Error on Patients, Family Members, and Health Professionals. In Riga, M. (Ed.), *Impact of Medical Errors and Malpractice on Health Economics, Quality, and Patient Safety* (pp. 171–196). Hershey, PA: IGI Global. DOI: 10.4018/978-1-5225-2337-6.ch007

Gündüz, U., & Pembecioğlu, N. (2021). Case Study of Gender and Career Choices: Being and Becoming Medical People. In G. Sarı (Ed.), *Handbook of Research on Representing Health and Medicine in Modern Media* (pp. 269-307). IGI Global. https://doi.org/10.4018/978-1-7998-6825-5.ch018

Gupta, A. (2021). Artificial Intelligence Approaches to Detect Neurodegenerative Disease From Medical Records: A Perspective. In Rani, G., & Tiwari, P. (Eds.), *Handbook of Research on Disease Prediction Through Data Analytics and Machine Learning* (pp. 254–267). IGI Global. https://doi.org/10.4018/978-1-7998-2742-9.ch013

Gürcü, M., & Tengilimoğlu, D. (2017). Health Tourism-Based Destination Marketing. In Bayraktar, A., & Uslay, C. (Eds.), *Strategic Place Branding Methodologies and Theory for Tourist Attraction* (pp. 308–331). Hershey, PA: IGI Global. DOI: 10.4018/978-1-5225-0579-2.ch015

Gürsel, G. (2017). For Better Healthcare Mining Health Data. In Bhattacharyya, S., De, S., Pan, I., & Dutta, P. (Eds.), *Intelligent Multidimensional Data Clustering and Analysis* (pp. 135–158). Hershey, PA: IGI Global. DOI: 10.4018/978-1-5225-1776-4.ch006

Gyaase, P. O., Darko-Lartey, R., William, H., & Borkloe, F. (2017). Towards an Integrated Electronic Medical Records System for Quality Healthcare in Ghana: An Exploratory Factor Analysis. *International Journal of Computers in Clinical Practice*, 2(2), 38–55. DOI: 10.4018/IJCCP.2017070103

Hagle, H. N., Liu, Y., Murphy, D. M., & Krom, L. (2022). Innovative Adaptation of Training and Technical Assistance: Education for the Behavioral Health Workforce. In Ford, C., & Garza, K. (Eds.), *Handbook of Research on Updating and Innovating Health Professions Education: Post-Pandemic Perspectives* (pp. 239–264). IGI Global. https://doi.org/10.4018/978-1-7998-7623-6.ch011

Halis, M., & Halis, M. (2021). The Relationship Between Organizational Culture and Organizational Commitment: A Research in Private Health Institutions. In G. Sarı (Ed.), *Handbook of Research on Representing Health and Medicine in Modern Media* (pp. 308-329). IGI Global. https://doi.org/10.4018/978-1-7998-6825-5.ch019

Halis, M., Tepret, S., Çamlibel, Z., & Halis, M. (2021). Competitiveness in Medical Tourism: An Evaluation on Kocaeli Medical Tourism Market. In G. Sarı (Ed.), *Handbook of Research on Representing Health and Medicine in Modern Media* (pp. 239-268). IGI Global. https://doi.org/10.4018/978-1-7998-6825-5.ch017

Hanel, P. (2017). Is China Catching Up?: Health-Related Applications of Biotechnology. In Bas, T., & Zhao, J. (Eds.), *Comparative Approaches to Biotechnology Development and Use in Developed and Emerging Nations* (pp. 465–520). Hershey, PA: IGI Global. DOI: 10.4018/978-1-5225-1040-6.ch016

Hartman, A., & Brown, S. (2017). Synergism through Therapeutic Visual Arts. In Bryan, V., & Bird, J. (Eds.), *Healthcare Community Synergism between Patients, Practitioners, and Researchers* (pp. 29–48). Hershey, PA: IGI Global. DOI: 10.4018/978-1-5225-0640-9.ch002

Heck, A. J., Cross, C. E., & Tatum, V. Y. (2020). Early Medical Education Readiness Interventions: Enhancing Undergraduate Preparedness. In Gotian, R., Kang, Y., & Safdieh, J. (Eds.), *Handbook of Research on the Efficacy of Training Programs and Systems in Medical Education* (pp. 283–304). IGI Global. https://doi.org/10.4018/978-1-7998-1468-9.ch015

Hemant, B., Arasappa, R. G. I., Udupa, K., & Varambally, S. (2021). Yoga for Mental Health Disorders: Research and Practice. In Telles, S., & Gupta, R. (Eds.), *Handbook of Research on Evidence-Based Perspectives on the Psychophysiology of Yoga and Its Applications* (pp. 179–198). IGI Global. https://doi.org/10.4018/978-1-7998-3254-6.ch011

Herdeiro, M. T., Gouveia, N., & Roque, F. (2021). Clinical Research and Regulatory Affairs: Skills and Tools in Pharmacy Education. In Figueiredo, I., & Cavaco, A. (Eds.), *Pedagogies for Pharmacy Curricula* (pp. 160–184). IGI Global. https://doi.org/10.4018/978-1-7998-4486-0.ch008

Hosoda, M., & Hosoda, M. (2019). Supporting Patient-led Initiatives to Improve Healthcare: An Investigation of Cancer and ME/CFS Support Groups and Prefectural Medical Councils in Japan. *International Journal of Patient-Centered Healthcare*, 9(2), 44–56. https://doi.org/10.4018/IJPCH.20190701.oa2

Iqbal, S., Ahmad, S., & Willis, I. (2017). Influencing Factors for Adopting Technology Enhanced Learning in the Medical Schools of Punjab, Pakistan. *International Journal of Information and Communication Technology Education*, 13(3), 27–39. DOI: 10.4018/IJICTE.2017070103

Jagiello, K., Sosnowska, A., Mikolajczyk, A., & Puzyn, T. (2017). Nanomaterials in Medical Devices: Regulations' Review and Future Perspectives. *Journal of Nanotoxicology and Nanomedicine*, 2(2), 1–11. DOI: 10.4018/JNN.2017070101

Jagtap, R., Phulare, K., Kurhade, M., & Gawande, K. S. (2021). Healthcare Conversational Chatbot for Medical Diagnosis. In Patil, B., & Vohra, M. (Eds.), *Handbook of Research on Engineering, Business, and Healthcare Applications of Data Science and Analytics* (pp. 401–415). IGI Global. https://doi.org/10.4018/978-1-7998-3053-5.ch020

Janosek, D. M. (2022). What Is the "Public Good" in a Pandemic? Who Decides?: Policy Makers and the Need for Leadership in Society's Perception of Medical Information. In Aker, H., & Aiken, M. (Eds.), *Handbook of Research on Cyberchondria, Health Literacy, and the Role of Media in Society's Perception of Medical Information* (pp. 1–15). IGI Global. https://doi.org/10.4018/978-1-7998-8630-3.ch001

Jena, T. K. (2018). Skill Training Process in Medicine Through Distance Mode. In Pandey, U., & Indrakanti, V. (Eds.), *Optimizing Open and Distance Learning in Higher Education Institutions* (pp. 228–243). Hershey, PA: IGI Global. DOI: 10.4018/978-1-5225-2624-7.ch010

Joseph, V., & Miller, J. M. (2018). Medical Students' Perceived Stigma in Seeking Care: A Cultural Perspective. In Smith, C. (Ed.), *Exploring the Pressures of Medical Education From a Mental Health and Wellness Perspective* (pp. 44–67). Hershey, PA: IGI Global. DOI: 10.4018/978-1-5225-2811-1.ch002

Joyce, B. L., & Swanberg, S. M. (2017). Using Backward Design for Competency-Based Undergraduate Medical Education. In Stefaniak, J. (Ed.), *Advancing Medical Education Through Strategic Instructional Design* (pp. 53–76). Hershey, PA: IGI Global. DOI: 10.4018/978-1-5225-2098-6.ch003

Kalder, M., & Kostev, K. (2021). Epidemiology of Gynaecological and Breast Cancers: Incidence, Survival, and Mental Health. In Dinas, K., Petousis, S., Kalder, M., & Mavromatidis, G. (Eds.), *Handbook of Research on Oncological and Endoscopical Dilemmas in Modern Gynecological Clinical Practice* (pp. 1–21). IGI Global. https://doi.org/10.4018/978-1-7998-4213-2.ch001

Kaljo, K., & Jacques, L. (2018). Flipping the Medical School Classroom. In M. Khosrow-Pour, D.B.A. (Ed.), *Encyclopedia of Information Science and Technology, Fourth Edition* (pp. 5800-5809). Hershey, PA: IGI Global. DOI: 10.4018/978-1-5225-2255-3.ch504

Karanfiloglu, M. (2021). The Role of Social Media Use in Health Communication: Digitized Health Communication During COVID-19 Pandemic. In G. Sarı (Eds.), *Handbook of Research on Representing Health and Medicine in Modern Media* (pp. 84-104). IGI Global. https://doi.org/10.4018/978-1-7998-6825-5.ch006

Karthikeyan, P., Vasuki, S., & Karthik, K. (2019). Non-Subsampled Contourlet Transform-Based Effective Denoising of Medical Images: Denoising of Medical Images Using Contourlet. In Swarnambiga, A. (Ed.), *Medical Image Processing for Improved Clinical Diagnosis* (pp. 155–183). IGI Global. https://doi.org/10.4018/978-1-5225-5876-7.ch008

Kashyap, R., & Rahamatkar, S. (2019). Medical Image Segmentation: An Advanced Approach. In Paul, S., Bhattacharya, P., & Bit, A. (Eds.), *Early Detection of Neurological Disorders Using Machine Learning Systems* (pp. 292–321). IGI Global. https://doi.org/10.4018/978-1-5225-8567-1.ch015

Kasina, H., Bahubalendruni, M. V., & Botcha, R. (2017). Robots in Medicine: Past, Present and Future. *International Journal of Manufacturing, Materials, and Mechanical Engineering*, 7(4), 44–64. DOI: 10.4018/IJMMME.2017100104

Katehakis, D. G. (2018). Electronic Medical Record Implementation Challenges for the National Health System in Greece. *International Journal of Reliable and Quality E-Healthcare*, 7(1), 16–30. DOI: 10.4018/IJRQEH.2018010102

Katehakis, D. G., & Kouroubali, A. (2022). Digital Transformation Challenges for the Development of Quality Electronic Medical Records in Greece. In Moumtzoglou, A. (Ed.), *Quality of Healthcare in the Aftermath of the COVID-19 Pandemic* (pp. 112–134). IGI Global. https://doi.org/10.4018/978-1-7998-9198-7.ch007

Kaur, P. D., & Sharma, P. (2017). Success Dimensions of ICTs in Healthcare. In Singh, B. (Ed.), *Computational Tools and Techniques for Biomedical Signal Processing* (pp. 149–173). Hershey, PA: IGI Global. DOI: 10.4018/978-1-5225-0660-7.ch007

Kaushik, P. (2018). Comorbidity of Medical and Psychiatric Disorders in Geriatric Population: Treatment and Care. In Prasad, B., & Akbar, S. (Eds.), *Handbook of Research on Geriatric Health, Treatment, and Care* (pp. 448–474). Hershey, PA: IGI Global. DOI: 10.4018/978-1-5225-3480-8.ch025

Khachane, M. Y. (2017). Organ-Based Medical Image Classification Using Support Vector Machine. *International Journal of Synthetic Emotions*, 8(1), 18–30. DOI: 10.4018/IJSE.2017010102

Kharbanda, V., Madaan, R., & Bhatia, K. K. (2021). Incautious Usage of Social Media: Impact on Emotional Intelligence and Health Concerns. In Yadav, D., Bansal, A., Bhatia, M., Hooda, M., & Morato, J. (Eds.), *Diagnostic Applications of Health Intelligence and Surveillance Systems* (pp. 172–186). IGI Global. https://doi.org/10.4018/978-1-7998-6527-8.ch008

Kirci, P. (2018). Intelligent Techniques for Analysis of Big Data About Healthcare and Medical Records. In Shah, N., & Mittal, M. (Eds.), *Handbook of Research on Promoting Business Process Improvement Through Inventory Control Techniques* (pp. 559–582). Hershey, PA: IGI Global. DOI: 10.4018/978-1-5225-3232-3.ch029

Kldiashvili, E. (2018). Cloud Approach for the Medical Information System: MIS on Cloud. In Khosrow-Pour, M. (Ed.), *Incorporating Nature-Inspired Paradigms in Computational Applications* (pp. 238–261). Hershey, PA: IGI Global. DOI: 10.4018/978-1-5225-5020-4.ch008

Ko, H., Mesicek, L., Choi, J., Choi, J., & Hwang, S. (2018). A Study on Secure Contents Strategies for Applications With DRM on Cloud Computing. *International Journal of Cloud Applications and Computing*, 8(1), 143–153. DOI: 10.4018/IJCAC.2018010107

Komendziński, T., Mikołajewska, E., & Mikołajewski, D. (2018). Cross-Cultural Decision-Making in Healthcare: Theory and Practical Application in Real Clinical Conditions. In Rosiek-Kryszewska, A., & Leksowski, K. (Eds.), *Healthcare Administration for Patient Safety and Engagement* (pp. 276–298). Hershey, PA: IGI Global. DOI: 10.4018/978-1-5225-3946-9.ch015

Konecny, L. T. (2018). Medical School Wellness Initiatives. In Smith, C. (Ed.), *Exploring the Pressures of Medical Education From a Mental Health and Wellness Perspective* (pp. 209–228). Hershey, PA: IGI Global. DOI: 10.4018/978-1-5225-2811-1.ch009

Kromrei, H., Solomonson, W. L., & Juzych, M. S. (2017). Teaching Residents How to Teach. In Stefaniak, J. (Ed.), *Advancing Medical Education Through Strategic Instructional Design* (pp. 164–185). Hershey, PA: IGI Global. DOI: 10.4018/978-1-5225-2098-6.ch008

Kulkarni, S., Savyanavar, A., Kulkarni, P., Stranieri, A., & Ghorpade, V. (2018). Framework for Integration of Medical Image and Text-Based Report Retrieval to Support Radiological Diagnosis. In Kolekar, M., & Kumar, V. (Eds.), *Biomedical Signal and Image Processing in Patient Care* (pp. 86–122). Hershey, PA: IGI Global. DOI: 10.4018/978-1-5225-2829-6.ch006

Kumar, A., & Sarkar, B. K. (2018). Performance Analysis of Nature-Inspired Algorithms-Based Bayesian Prediction Models for Medical Data Sets. In Singh, U., Tiwari, A., & Singh, R. (Eds.), *Soft-Computing-Based Nonlinear Control Systems Design* (pp. 134–155). Hershey, PA: IGI Global. DOI: 10.4018/978-1-5225-3531-7.ch007

Labbadi, W., & Akaichi, J. (2017). Efficient Algorithm for Answering Fuzzy Medical Requests in Pervasive Healthcare Information Systems. *International Journal of Healthcare Information Systems and Informatics*, 12(2), 46–64. DOI: 10.4018/IJHISI.2017040103

Lagumdzija, A., & Swing, V. K. (2017). Health, Digitalization, and Individual Empowerment. In Topor, F. (Ed.), *Handbook of Research on Individualism and Identity in the Globalized Digital Age* (pp. 380–402). Hershey, PA: IGI Global. DOI: 10.4018/978-1-5225-0522-8.ch017

Lamey, T. W., & Davidson-Shivers, G. V. (2017). Instructional Strategies and Sequencing. In Stefaniak, J. (Ed.), *Advancing Medical Education Through Strategic Instructional Design* (pp. 30–52). Hershey, PA: IGI Global. DOI: 10.4018/978-1-5225-2098-6.ch002

Lee, D. C., & Gefen, D. (2020). Promises and Challenges of Medical Patient Healthcare Portals in Underserved Communities: The Case of Einstein Medical Center Philadelphia (EMCP). In McHaney, R., Reychev, I., Azuri, J., McHaney, M., & Moshonov, R. (Eds.), *Impacts of Information Technology on Patient Care and Empowerment* (pp. 219–251). IGI Global. https://doi.org/10.4018/978-1-7998-0047-7.ch012

Leon, G. (2017). The Role of Forensic Medicine in Medical Errors. In Riga, M. (Ed.), *Impact of Medical Errors and Malpractice on Health Economics, Quality, and Patient Safety* (pp. 144–170). Hershey, PA: IGI Global. DOI: 10.4018/978-1-5225-2337-6.ch006

Leung, L. Y. (2020). Representation of Gender Equality From the Perspective of the Medical Trainee and its Ripple Effect: Highlighting Gender Inequality in Medical Student Experiences. In Bellini, M., & Papalois, V. (Eds.), *Gender Equity in the Medical Profession* (pp. 29–50). IGI Global. https://doi.org/10.4018/978-1-5225-9599-1.ch003

Lopes, M. J., Fonseca, C., & Barbosa, P. (2020). The Individual Care Plan as Electronic Health Record: A Tool for Management, Integration of Care, and Better Health Results. In Mendes, D., Fonseca, C., Lopes, M., García-Alonso, J., & Murillo, J. (Eds.), *Exploring the Role of ICTs in Healthy Aging* (pp. 1–12). IGI Global. https://doi.org/10.4018/978-1-7998-1937-0.ch001

Love, L., & McDowelle, D. (2018). Developing a Comprehensive Wellness Program for Medical Students. In Smith, C. (Ed.), *Exploring the Pressures of Medical Education From a Mental Health and Wellness Perspective* (pp. 190–208). Hershey, PA: IGI Global. DOI: 10.4018/978-1-5225-2811-1.ch008

Lovell, K. L. (2017). Development and Evaluation of Neuroscience Computer-Based Modules for Medical Students: Instructional Design Principles and Effectiveness. In Stefaniak, J. (Ed.), *Advancing Medical Education Through Strategic Instructional Design* (pp. 262–276). Hershey, PA: IGI Global. DOI: 10.4018/978-1-5225-2098-6.ch013

Lubin, R., & Hamlin, M. D. (2018). Medical Student Burnout: A Social Cognitive Learning Perspective on Medical Student Mental Health and Wellness. In Smith, C. (Ed.), *Exploring the Pressures of Medical Education From a Mental Health and Wellness Perspective* (pp. 92–121). Hershey, PA: IGI Global. DOI: 10.4018/978-1-5225-2811-1.ch004

Luk, C. Y. (2018). Moving Towards Universal Health Coverage: Challenges for the Present and Future in China. In Fong, B., Ng, A., & Yuen, P. (Eds.), *Sustainable Health and Long-Term Care Solutions for an Aging Population* (pp. 19–45). Hershey, PA: IGI Global. DOI: 10.4018/978-1-5225-2633-9.ch002

Mackintosh, S. E., & Katsaros, E. (2020). Building Interprofessional Competencies Into Medical Education and Assessment. In Waldman, S., & Bowlin, S. (Eds.), *Building a Patient-Centered Interprofessional Education Program* (pp. 84–112). IGI Global. https://doi.org/10.4018/978-1-7998-3066-5.ch005

Magalhães, J. L., Quoniam, L., Hartz, Z., Silveira, H., & Rito, P. D. (2022). Knowledge Management in Big Data Times for Global Health: Challenges for Quality in One Health. In Lima de Magalhães, J., Hartz, Z., Jamil, G., Silveira, H., & Jamil, L. (Eds.), *Handbook of Research on Essential Information Approaches to Aiding Global Health in the One Health Context* (pp. 149–163). IGI Global. https://doi.org/10.4018/978-1-7998-8011-0.ch008

Mahat, M., & Pettigrew, A. (2017). The Regulatory Environment of Non-Profit Higher Education and Research Institutions and Its Implications for Managerial Strategy: The Case of Medical Education and Research. In West, L., & Worthington, A. (Eds.), *Handbook of Research on Emerging Business Models and Managerial Strategies in the Nonprofit Sector* (pp. 336–351). Hershey, PA: IGI Global. DOI: 10.4018/978-1-5225-2537-0.ch017

Maitra, I. K., & Bandhyopadhyaay, S. K. (2017). Adaptive Edge Detection Method towards Features Extraction from Diverse Medical Imaging Technologies. In Bhattacharyya, S., De, S., Pan, I., & Dutta, P. (Eds.), *Intelligent Multidimensional Data Clustering and Analysis* (pp. 159–192). Hershey, PA: IGI Global. DOI: 10.4018/978-1-5225-1776-4.ch007

Mane, V. M., Patki, S., Dhumma, A. V., & Apte, K. J. (2022). Telehealth System for Effective Treatment in COVID-19 Pandemics. In Thakare, A., Wagh, S., Bhende, M., Anter, A., & Gao, X. (Eds.), *Handbook of Research on Applied Intelligence for Health and Clinical Informatics* (pp. 328–339). IGI Global. https://doi.org/10.4018/978-1-7998-7709-7.ch019

Mangu, V. P. (2017). Mobile Health Care: A Technology View. In Bhatt, C., & Peddoju, S. (Eds.), *Cloud Computing Systems and Applications in Healthcare* (pp. 1–18). Hershey, PA: IGI Global. DOI: 10.4018/978-1-5225-1002-4.ch001

Manirabona, A., Fourati, L. C., & Boudjit, S. (2017). Investigation on Healthcare Monitoring Systems: Innovative Services and Applications. *International Journal of E-Health and Medical Communications*, 8(1), 1–18. DOI: 10.4018/IJEHMC.2017010101

Martins, C. L., Martinho, D., Marreiros, G., Conceição, L., Faria, L., & Simões de Almeida, R. (2022). Artificial Intelligence in Digital Mental Health. In Marques, A., & Queirós, R. (Eds.), *Digital Therapies in Psychosocial Rehabilitation and Mental Health* (pp. 201–225). IGI Global. https://doi.org/10.4018/978-1-7998-8634-1.ch010

Marwan, M., Kartit, A., & Ouahmane, H. (2018). A Framework to Secure Medical Image Storage in Cloud Computing Environment. *Journal of Electronic Commerce in Organizations*, 16(1), 1–16. DOI: 10.4018/JECO.2018010101

Mason, A., Bhati, S., Jiang, R., & Spencer, E. A. (2020). Medical Tourism Patient Mortality: Considerations From a 10-Year Review of Global News Media Representations. In Merviö, M. (Ed.), *Global Issues and Innovative Solutions in Healthcare, Culture, and the Environment* (pp. 206–225). IGI Global. https://doi.org/10.4018/978-1-7998-3576-9.ch011

Masoud, M. P., Nejad, M. K., Darebaghi, H., Chavoshi, M., & Farahani, M. (2018). The Decision Support System and Conventional Method of Telephone Triage by Nurses in Emergency Medical Services: A Comparative Investigation. *International Journal of E-Business Research*, 14(1), 77–88. DOI: 10.4018/IJEBR.2018010105

Mathew, N. E., Walter, A., & Eapen, V. (2020). Mental Health Challenges in Children With Intellectual Disabilities. In Gopalan, R. (Ed.), *Developmental Challenges and Societal Issues for Individuals With Intellectual Disabilities* (pp. 13–39). IGI Global. https://doi.org/10.4018/978-1-7998-1223-4.ch002

Mazzola, A. (2020). The Social Mandate to Deal With Mental Health: A Comparison Between Interventions in a Mental Health Center, a School, and a Psychoanalytic Office. In Padmanaban, S., & Subudhi, C. (Eds.), *Psycho-Social Perspectives on Mental Health and Well-Being* (pp. 234–254). IGI Global. https://doi.org/10.4018/978-1-7998-1185-5.ch012

McDonald, W. G., Martin, M., & Salzberg, L. D. (2018). From Medical Student to Medical Resident: Graduate Medical Education and Mental Health in the United States. In Smith, C. (Ed.), *Exploring the Pressures of Medical Education From a Mental Health and Wellness Perspective* (pp. 145–169). Hershey, PA: IGI Global. DOI: 10.4018/978-1-5225-2811-1.ch006

McGrowder, D. A., Miller, F. G., Nwokocha, C., Wilson-Clarke, C. F., Anderson, M., Anderson-Jackson, L., Williams, L., & Alexander-Lindo, R. (2021). Medical Herbs and the Treatment of Diabetes Mellitus: Mechanisms of Action. In Hussain, A., & Behl, S. (Eds.), *Treating Endocrine and Metabolic Disorders With Herbal Medicines* (pp. 48–73). IGI Global. https://doi.org/10.4018/978-1-7998-4808-0.ch003

Medhekar, A. (2017). The Role of Social Media for Knowledge Dissemination in Medical Tourism: A Case of India. In Chugh, R. (Ed.), *Harnessing Social Media as a Knowledge Management Tool* (pp. 25–54). Hershey, PA: IGI Global. DOI: 10.4018/978-1-5225-0495-5.ch002

Medhekar, A. (2020). Digital Health Innovation Enhancing Patient Experience in Medical Travel. In Sandhu, K. (Ed.), *Opportunities and Challenges in Digital Healthcare Innovation* (pp. 13–35). IGI Global. https://doi.org/10.4018/978-1-7998-3274-4.ch002

Medhekar, A., & Haq, F. (2018). Urbanization and New Jobs Creation in Healthcare Services in India: Challenges and Opportunities. In Benna, U., & Benna, I. (Eds.), *Urbanization and Its Impact on Socio-Economic Growth in Developing Regions* (pp. 198–218). Hershey, PA: IGI Global. DOI: 10.4018/978-1-5225-2659-9.ch010

Mehta, P. (2017). Framework of Indian Healthcare System and its Challenges: An Insight. In Bryan, V., & Bird, J. (Eds.), *Healthcare Community Synergism between Patients, Practitioners, and Researchers* (pp. 247–271). Hershey, PA: IGI Global. DOI: 10.4018/978-1-5225-0640-9.ch011

Mendes, E., Sousa, B. B., & Gonçalves, M. (2021). The Role of Technologies in Relationship Management and Internal Marketing: An Approach in the Health Sector. In J. Santos, & I. Pereira (Ed.), *Management and Marketing for Improved Competitiveness and Performance in the Healthcare Sector* (pp. 213-237). IGI Global. https://doi.org/10.4018/978-1-7998-7263-4.ch010

Meyer, T. M., & Simen, J. H. (2020). Geriatric Interprofessional Education: A Practical Guide for Health Professions Educators. In Waldman, S., & Bowlin, S. (Eds.), *Building a Patient-Centered Interprofessional Education Program* (pp. 120–136). IGI Global. https://doi.org/10.4018/978-1-7998-3066-5.ch007

Mi, M. (2017). Informal Learning in Medical Education. In Stefaniak, J. (Ed.), *Advancing Medical Education Through Strategic Instructional Design* (pp. 225–244). Hershey, PA: IGI Global. DOI: 10.4018/978-1-5225-2098-6.ch011

Mishra, K., & Thomas, R. (2022). Low Cost, User-Controlled Peroneal Stimulator for Foot Drop in Patients With Stroke: An Experiment in Indian Rehabilitation Set-Up. In Stasolla, F. (Ed.), *Assistive Technologies for Assessment and Recovery of Neurological Impairments* (pp. 279–303). IGI Global. https://doi.org/10.4018/978-1-7998-7430-0.ch014

Mishra, S., & Panda, M. (2019). Artificial Intelligence in Medical Science. In Bouchemal, N. (Ed.), *Intelligent Systems for Healthcare Management and Delivery* (pp. 306–330). IGI Global. https://doi.org/10.4018/978-1-5225-7071-4.ch014

Mosadeghrad, A. M., & Woldemichael, A. (2017). Application of Quality Management in Promoting Patient Safety and Preventing Medical Errors. In Riga, M. (Ed.), *Impact of Medical Errors and Malpractice on Health Economics, Quality, and Patient Safety* (pp. 91–112). Hershey, PA: IGI Global. DOI: 10.4018/978-1-5225-2337-6.ch004

Moumtzoglou, A. (2017). Digital Medicine: The Quality Standpoint. In Moumtzoglou, A. (Ed.), *Design, Development, and Integration of Reliable Electronic Healthcare Platforms* (pp. 179–195). Hershey, PA: IGI Global. DOI: 10.4018/978-1-5225-1724-5.ch011

Moumtzoglou, A., & Pouliakis, A. (2018). Population Health Management and the Science of Individuality. *International Journal of Reliable and Quality E-Healthcare*, 7(2), 1–26. DOI: 10.4018/IJRQEH.2018040101

Mukhtar, W. F., & Abuelyaman, E. S. (2017). Opportunities and Challenges of Big Data in Healthcare. In Wickramasinghe, N. (Ed.), *Handbook of Research on Healthcare Administration and Management* (pp. 47–58). Hershey, PA: IGI Global. DOI: 10.4018/978-1-5225-0920-2.ch004

Munugala, S., Brar, G. K., Syed, A., Mohammad, A., & Halgamuge, M. N. (2018). The Much Needed Security and Data Reforms of Cloud Computing in Medical Data Storage. In Lytras, M., & Papadopoulou, P. (Eds.), *Applying Big Data Analytics in Bioinformatics and Medicine* (pp. 99–113). Hershey, PA: IGI Global. DOI: 10.4018/978-1-5225-2607-0.ch005

Naha, N., & Manickavachagam, G. (2021). Textile Industry and Health Hazards: Impact of Climate Change Issues and Fertility Potential. In K. Wani, & N. Naha (Ed.), *Climate Change and Its Impact on Fertility* (pp. 42-69). IGI Global. https://doi.org/10.4018/978-1-7998-4480-8.ch003

Nair, L. D. (2019). Childhood Neurodegenerative Disorders. In Uddin, M., & Amran, M. (Eds.), *Handbook of Research on Critical Examinations of Neurodegenerative Disorders* (pp. 385–412). IGI Global. https://doi.org/10.4018/978-1-5225-5282-6.ch018

Nayak, S. R., & Mishra, J. (2019). Analysis of Medical Images Using Fractal Geometry. In Dey, N., Ashour, A., Kalia, H., Goswami, R., & Das, H. (Eds.), *Histopathological Image Analysis in Medical Decision Making* (pp. 181–201). IGI Global. https://doi.org/10.4018/978-1-5225-6316-7.ch008

Netto, J. T., Hartz, Z., & de Magalhães, J. L. (2022). One Health and Information Management Using Big Data in Health: A Brazilian Case Study for COVID-19. In Lima de Magalhães, J., Hartz, Z., Jamil, G., Silveira, H., & Jamil, L. (Eds.), *Handbook of Research on Essential Information Approaches to Aiding Global Health in the One Health Context* (pp. 36–51). IGI Global. https://doi.org/10.4018/978-1-7998-8011-0.ch003

Ngara, R. (2017). Multiple Voices, Multiple Paths: Towards Dialogue between Western and Indigenous Medical Knowledge Systems. In Ngulube, P. (Ed.), *Handbook of Research on Theoretical Perspectives on Indigenous Knowledge Systems in Developing Countries* (pp. 332–358). Hershey, PA: IGI Global. DOI: 10.4018/978-1-5225-0833-5.ch015

Nikumbh, D. D., Sayyad, S., Joshi, R. R., Dubey, K. S., Mehta, D. V., & Matta, D. K. (2022). Applied Intelligence for Medical Diagnosing. In Thakare, A., Wagh, S., Bhende, M., Anter, A., & Gao, X. (Eds.), *Handbook of Research on Applied Intelligence for Health and Clinical Informatics* (pp. 44–79). IGI Global. https://doi.org/10.4018/978-1-7998-7709-7.ch004

O'Connor, Y., & Heavin, C. (2018). Defining and Characterising the Landscape of eHealth. In M. Khosrow-Pour, D.B.A. (Ed.), *Encyclopedia of Information Science and Technology, Fourth Edition* (pp. 5864-5875). Hershey, PA: IGI Global. DOI: 10.4018/978-1-5225-2255-3.ch510

Ochonogor, W. C., & Okite-Amughoro, F. A. (2018). Building an Effective Digital Library in a University Teaching Hospital (UTH) in Nigeria. In Tella, A., & Kwanya, T. (Eds.), *Handbook of Research on Managing Intellectual Property in Digital Libraries* (pp. 184–204). Hershey, PA: IGI Global. DOI: 10.4018/978-1-5225-3093-0.ch010

Olivares, S. L., Cruz, A. G., Cabrera, M. V., Regalado, A. I., & García, J. E. (2017). An Assessment Study of Quality Model for Medical Schools in Mexico. In Mukerji, S., & Tripathi, P. (Eds.), *Handbook of Research on Administration, Policy, and Leadership in Higher Education* (pp. 404–439). Hershey, PA: IGI Global. DOI: 10.4018/978-1-5225-0672-0.ch016

Omoruyi, E. A., & Omidele, F. (2018). Resident Physician and Medical Academic Faculty Burnout: A Review of Current Literature. In Smith, C. (Ed.), *Exploring the Pressures of Medical Education From a Mental Health and Wellness Perspective* (pp. 171–189). Hershey, PA: IGI Global. DOI: 10.4018/978-1-5225-2811-1.ch007

Pandey, A., Singh, B., Saini, B. S., & Sood, N. (2019). Medical Data Security Tools and Techniques in E-Health Applications. In Singh, B., Saini, B., Singh, D., & Pandey, A. (Eds.), *Medical Data Security for Bioengineers* (pp. 124–131). IGI Global. https://doi.org/10.4018/978-1-5225-7952-6.ch006

Papadopoulou, P., Lytras, M., & Marouli, C. (2018). Bioinformatics as Applied to Medicine: Challenges Faced Moving from Big Data to Smart Data to Wise Data. In Lytras, M., & Papadopoulou, P. (Eds.), *Applying Big Data Analytics in Bioinformatics and Medicine* (pp. 1–25). Hershey, PA: IGI Global. DOI: 10.4018/978-1-5225-2607-0.ch001

Paraskou, A., & George, B. P. (2018). An Overview of Reproductive Tourism. In *Legal and Economic Considerations Surrounding Reproductive Tourism: Emerging Research and Opportunities* (pp. 1–17). Hershey, PA: IGI Global. DOI: 10.4018/978-1-5225-2694-0.ch001

Pereira, S., Silva, L., Machado, J., & Cabral, A. (2020). The Clinical Informatization in Portugal: An Approach to the National Health Service Certification. *International Journal of Reliable and Quality E-Healthcare*, 9(2), 34–47. https://doi.org/10.4018/IJRQEH.2020040103

Phuritsabam, B., & Devi, A. B. (2017). Information Seeking Behavior of Medical Scientists at Jawaharlal Nehru Institute of Medical Science: A Study. In Ram, S. (Ed.), *Library and Information Services for Bioinformatics Education and Research* (pp. 177–187). Hershey, PA: IGI Global. DOI: 10.4018/978-1-5225-1871-6.ch010

Pieczka, B. (2018). Management of Risk and Adverse Events in Medical Entities. In Rosiek-Kryszewska, A., & Leksowski, K. (Eds.), *Healthcare Administration for Patient Safety and Engagement* (pp. 31–46). Hershey, PA: IGI Global. DOI: 10.4018/978-1-5225-3946-9.ch003

Poduval, J. (2017). Medical Errors: Impact on Health Care Quality. In Riga, M. (Ed.), *Impact of Medical Errors and Malpractice on Health Economics, Quality, and Patient Safety* (pp. 33–60). Hershey, PA: IGI Global. DOI: 10.4018/978-1-5225-2337-6.ch002

Politis, D., Stagiopoulos, P., Aidona, S., Kyriafinis, G., & Constantinidis, I. (2018). Autonomous Learning and Skill Accreditation: A Paradigm for Medical Studies. In Kumar, A. (Ed.), *Optimizing Student Engagement in Online Learning Environments* (pp. 266–296). Hershey, PA: IGI Global. DOI: 10.4018/978-1-5225-3634-5.ch012

Pomares-Quimbaya, A., González, R. A., Sierra, A., Daza, J. C., Muñoz, O., García, A., & Bohórquez, W. R. (2017). ICT for Enabling the Quality Evaluation of Health Care Services: A Case Study in a General Hospital. In Moumtzoglou, A. (Ed.), *Design, Development, and Integration of Reliable Electronic Healthcare Platforms* (pp. 196–210). Hershey, PA: IGI Global. DOI: 10.4018/978-1-5225-1724-5.ch012

Ponnuswamy, P., Paul, A. S., & Jose, A. B. (2021). Pre-, Peri-, and Post-Natal Risk Factors in ADHD. In Gopalan, R. (Ed.), *New Developments in Diagnosing, Assessing, and Treating ADHD* (pp. 43–54). IGI Global. https://doi.org/10.4018/978-1-7998-5495-1.ch003

Pouliakis, A., Margari, N., Karakitsou, E., Archondakis, S., & Karakitsos, P. (2018). Emerging Technologies Serving Cytopathology: Big Data, the Cloud, and Mobile Computing. In El Naqa, I. (Ed.), *Emerging Developments and Practices in Oncology* (pp. 114–152). Hershey, PA: IGI Global. DOI: 10.4018/978-1-5225-3085-5.ch005

Praveenkumar, P., Santhiyadevi, R., & Amirtharajan, R. (2019). Medical Data Are Safe: An Encrypted Quantum Approach. In Singh, B., Saini, B., Singh, D., & Pandey, A. (Eds.), *Medical Data Security for Bioengineers* (pp. 142–165). IGI Global. https://doi.org/10.4018/978-1-5225-7952-6.ch008

Punhani, R., Saini, S., Varun, N., & Rustagi, R. (2021). mHealth: A Resolution in Improving Global Health. In D. Yadav, A. Bansal, M. Bhatia, M. Hooda, & J. Morato (Ed.), *Diagnostic Applications of Health Intelligence and Surveillance Systems* (pp. 86-105). IGI Global. https://doi.org/10.4018/978-1-7998-6527-8.ch004

Quintela, A., & Veiga, A. (2022). An Integrated Approach (Social and Health) in the Elderly With Diabetes: An Increase in Health Literacy. In Vaz de Almeida, C., & Ramos, S. (Eds.), *Handbook of Research on Assertiveness, Clarity, and Positivity in Health Literacy* (pp. 134–142). IGI Global. https://doi.org/10.4018/978-1-7998-8824-6.ch008

Rai, A., Kothari, R., & Singh, D. P. (2017). Assessment of Available Technologies for Hospital Waste Management: A Need for Society. In Singh, R., Singh, A., & Srivastava, V. (Eds.), *Environmental Issues Surrounding Human Overpopulation* (pp. 172–188). Hershey, PA: IGI Global. DOI: 10.4018/978-1-5225-1683-5.ch010

Rajagopalan, S., Janakiraman, S., & Rengarajan, A. (2019). Medical Image Encryption: Microcontroller and FPGA Perspective. In Singh, B., Saini, B., Singh, D., & Pandey, A. (Eds.), *Medical Data Security for Bioengineers* (pp. 278–304). IGI Global. https://doi.org/10.4018/978-1-5225-7952-6.ch014

Ramamoorthy, S., & Sivasubramaniam, R. (2018). Image Processing Including Medical Liver Imaging: Medical Image Processing from Big Data Perspective, Ultrasound Liver Images, Challenges. In Lytras, M., & Papadopoulou, P. (Eds.), *Applying Big Data Analytics in Bioinformatics and Medicine* (pp. 380–392). Hershey, PA: IGI Global. DOI: 10.4018/978-1-5225-2607-0.ch016

Rathor, G. P., & Gupta, S. K. (2017). Improving Multimodality Image Fusion through Integrate AFL and Wavelet Transform. In Tiwari, V., Tiwari, B., Thakur, R., & Gupta, S. (Eds.), *Pattern and Data Analysis in Healthcare Settings* (pp. 143–157). Hershey, PA: IGI Global. DOI: 10.4018/978-1-5225-0536-5.ch008

Rexhepi, H., & Persson, A. (2017). Challenges to Implementing IT Support for Evidence Based Practice Among Nurses and Assistant Nurses: A Qualitative Study. *Journal of Electronic Commerce in Organizations*, 15(2), 61–76. DOI: 10.4018/JECO.2017040105

Ribeiro, A. D., Ribeiro, L. P., Silva, C. A., & Magalhães-Ribeiro, L. P. (2021). Patient Safety: The Patient's Perspective From Public Sector Health Institutions in the Algarve Region, Portugal. In Eklund, P. (Ed.), *Integrated Care and Fall Prevention in Active and Healthy Aging* (pp. 200–222). IGI Global. https://doi.org/10.4018/978-1-7998-4411-2.ch011

Rodrigues, H., & Brochado, A. (2021). Going for Silver-Senior Consumers' Reviews of Medical Tourism. In Borges, A., & Rodrigues, P. (Eds.), *New Techniques for Brand Management in the Healthcare Sector* (pp. 64–83). IGI Global. https://doi.org/10.4018/978-1-7998-3034-4.ch005

Rodrigues, J. M., Oliveira, F., Ribeiro, C. P., & Santos, R. C. (2022). Mobile Mental Health for Depression Assistance: Research Directions, Obstacles, Advantages, and Disadvantages of Implementing mHealth. In Marques, A., & Queirós, R. (Eds.), *Digital Therapies in Psychosocial Rehabilitation and Mental Health* (pp. 21–40). IGI Global. https://doi.org/10.4018/978-1-7998-8634-1.ch002

Ros, M., Weaver, L., & Neuwirth, L. S. (2020). Virtual Reality Stereoscopic 180-Degree Video-Based Immersive Environments: Applications for Training Surgeons and Other Medical Professionals. In Stefaniak, J. (Ed.), *Cases on Instructional Design and Performance Outcomes in Medical Education* (pp. 92–119). IGI Global. https://doi.org/10.4018/978-1-7998-5092-2.ch005

Rosiek, A. (2018). The Assessment of Actions of the Environment and the Impact of Preventive Medicine for Public Health in Poland. In Rosiek-Kryszewska, A., & Leksowski, K. (Eds.), *Healthcare Administration for Patient Safety and Engagement* (pp. 106–119). Hershey, PA: IGI Global. DOI: 10.4018/978-1-5225-3946-9.ch006

Rosiek, A., & Rosiek-Kryszewska, A. (2018). Managed Healthcare: Doctor Life Satisfaction and Its Impact on the Process of Communicating With the Patient. In Rosiek-Kryszewska, A., & Leksowski, K. (Eds.), *Healthcare Administration for Patient Safety and Engagement* (pp. 244–261). Hershey, PA: IGI Global. DOI: 10.4018/978-1-5225-3946-9.ch013

Rosiek-Kryszewska, A., & Rosiek, A. (2018). The Involvement of the Patient and his Perspective Evaluation of the Quality of Healthcare. In Rosiek-Kryszewska, A., & Leksowski, K. (Eds.), *Healthcare Administration for Patient Safety and Engagement* (pp. 121–144). Hershey, PA: IGI Global. DOI: 10.4018/978-1-5225-3946-9.ch007

Rosiek-Kryszewska, A., & Rosiek, A. (2018). The Impact of Management and Leadership Roles in Building Competitive Healthcare Units. In Rosiek-Kryszewska, A., & Leksowski, K. (Eds.), *Healthcare Administration for Patient Safety and Engagement* (pp. 13–30). Hershey, PA: IGI Global. DOI: 10.4018/978-1-5225-3946-9.ch002

Rouzbehani, K. (2017). Health Policy Implementation: Moving Beyond Its Barriers in United States. In Wickramasinghe, N. (Ed.), *Handbook of Research on Healthcare Administration and Management* (pp. 541–552). Hershey, PA: IGI Global. DOI: 10.4018/978-1-5225-0920-2.ch032

Sajjad, N., Hassan, S., Qadir, J., Ali, R., & Shah, D. (2020). Analyses of the Recycling Potential of Medical Plastic Wastes. In Wani, K., Ariana, L., & Zuber, S. (Eds.), *Handbook of Research on Environmental and Human Health Impacts of Plastic Pollution* (pp. 178–199). IGI Global. https://doi.org/10.4018/978-1-5225-9452-9.ch010

Sam, S. (2017). Mobile Phones and Expanding Human Capabilities in Plural Health Systems. In Moahi, K., Bwalya, K., & Sebina, P. (Eds.), *Health Information Systems and the Advancement of Medical Practice in Developing Countries* (pp. 93–114). Hershey, PA: IGI Global. DOI: 10.4018/978-1-5225-2262-1.ch006

Santos-Trigo, M., Suaste, E., & Figuerola, P. (2018). Technology Design and Routes for Tool Appropriation in Medical Practices. In M. Khosrow-Pour, D.B.A. (Ed.), *Encyclopedia of Information Science and Technology, Fourth Edition* (pp. 3794-3804). Hershey, PA: IGI Global. DOI: 10.4018/978-1-5225-2255-3.ch329

Sarivougioukas, J., Vagelatos, A., Parsopoulos, K. E., & Lagaris, I. E. (2018). Home UbiHealth. In M. Khosrow-Pour, D.B.A. (Ed.), *Encyclopedia of Information Science and Technology, Fourth Edition* (pp. 7765-7774). Hershey, PA: IGI Global. DOI: 10.4018/978-1-5225-2255-3.ch675

Sarkar, B. K. (2017). Big Data and Healthcare Data: A Survey. *International Journal of Knowledge-Based Organizations*, 7(4), 50–77. DOI: 10.4018/IJKBO.2017100104

Sarkar, K., & Li, B. (2021). Deep Learning for Medical Image Segmentation. In Saxena, S., & Paul, S. (Eds.), *Deep Learning Applications in Medical Imaging* (pp. 40–77). IGI Global. https://doi.org/10.4018/978-1-7998-5071-7.ch002

Savani, M. M. (2021). Commitment Devices for Health: Theory and Evidence on Weight Loss. In Mihaila, V. (Ed.), *Behavioral-Based Interventions for Improving Public Policies* (pp. 35–54). IGI Global. https://doi.org/10.4018/978-1-7998-2731-3.ch003

Saxena, K., & Banodha, U. (2017). An Essence of the SOA on Healthcare. In Bhadoria, R., Chaudhari, N., Tomar, G., & Singh, S. (Eds.), *Exploring Enterprise Service Bus in the Service-Oriented Architecture Paradigm* (pp. 283–304). Hershey, PA: IGI Global. DOI: 10.4018/978-1-5225-2157-0.ch018

Saxena, N. (2020). Considerations for Stakeholders of Medical Tourism: A Comprehensive Examination. In Paul, S., & Kulshreshtha, S. (Eds.), *Global Developments in Healthcare and Medical Tourism* (pp. 23–37). IGI Global. https://doi.org/10.4018/978-1-5225-9787-2.ch002

Segura-Azuara, N. D., Guzman-Segura, J. G., Guzmán-Segura, N. M., & Guzmán-Segura, J. P. (2022). Disease Awareness Campaigns: Education for Citizenship in Medical Schools. In Lopez, M. (Ed.), *Advancing Health Education With Telemedicine* (pp. 113–122). IGI Global. https://doi.org/10.4018/978-1-7998-8783-6.ch006

Sekaran, P. A. (2022). Smart Medical Kit in Chronic Kidney Disease Management. In Jeya Mala, D. (Ed.), *Integrating AI in IoT Analytics on the Cloud for Healthcare Applications* (pp. 24–40). IGI Global. https://doi.org/10.4018/978-1-7998-9132-1.ch002

Sen, K., & Ghosh, K. (2018). Incorporating Global Medical Knowledge to Solve Healthcare Problems: A Framework for a Crowdsourcing System. *International Journal of Healthcare Information Systems and Informatics*, 13(1), 1–14. DOI: 10.4018/IJHISI.2018010101

Senanayake, B., Tyagi, N., Zhou, X., & Edirippulige, S. (2020). Workforce Readiness and Digital Health Integration. In Sandhu, K. (Ed.), *Opportunities and Challenges in Digital Healthcare Innovation* (pp. 170–185). IGI Global. https://doi.org/10.4018/978-1-7998-3274-4.ch010

Shakdher, A., & Pandey, K. (2017). REDAlert+: Medical/Fire Emergency and Warning System using Android Devices. *International Journal of E-Health and Medical Communications*, 8(1), 37–51. DOI: 10.4018/IJEHMC.2017010103

Shekarian, E., Abdul-Rashid, S. H., & Olugu, E. U. (2017). An Integrated Fuzzy VIKOR Method for Performance Management in Healthcare. In Tavana, M., Szabat, K., & Puranam, K. (Eds.), *Organizational Productivity and Performance Measurements Using Predictive Modeling and Analytics* (pp. 40–61). Hershey, PA: IGI Global. DOI: 10.4018/978-1-5225-0654-6.ch003

Shijina, V., & John, S. J. (2017). Multiple Relations and its Application in Medical Diagnosis. *International Journal of Fuzzy System Applications*, 6(4), 47–62. DOI: 10.4018/IJFSA.2017100104

Shipley, N., & Chakraborty, J. (2018). Big Data and mHealth: Increasing the Usability of Healthcare Through the Customization of Pinterest – Literary Perspective. In Machado, J., Abelha, A., Santos, M., & Portela, F. (Eds.), *Next-Generation Mobile and Pervasive Healthcare Solutions* (pp. 46–66). Hershey, PA: IGI Global. DOI: 10.4018/978-1-5225-2851-7.ch004

Shridevi, S., & Saleena, B. (2019). Semantic Technologies for Medical Knowledge Representation. In Puratchikody, A., Prabu, S., & Umamaheswari, A. (Eds.), *Computer Applications in Drug Discovery and Development* (Vol. 5, pp. 260–275). IGI Global. https://doi.org/10.4018/978-1-5225-7326-5.ch012

Simões de Almeida, R. (2022). Mobile Mental Health: Opportunities and Challenges. In Marques, A., & Queirós, R. (Eds.), *Digital Therapies in Psychosocial Rehabilitation and Mental Health* (pp. 1–20). IGI Global. https://doi.org/10.4018/978-1-7998-8634-1.ch001

Simões de Almeida, R., & da Silva, T. P. (2022). AI Chatbots in Mental Health: Are We There Yet? In Marques, A., & Queirós, R. (Eds.), *Digital Therapies in Psychosocial Rehabilitation and Mental Health* (pp. 226–243). IGI Global. https://doi.org/10.4018/978-1-7998-8634-1.ch011

Singh, A., & Dutta, M. K. (2017). A Reversible Data Hiding Scheme for Efficient Management of Tele-Ophthalmological Data. *International Journal of E-Health and Medical Communications*, 8(3), 38–54. DOI: 10.4018/IJEHMC.2017070103

Skourti, P. K., & Pavlakis, A. (2017). The Second Victim Phenomenon: The Way Out. In Riga, M. (Ed.), *Impact of Medical Errors and Malpractice on Health Economics, Quality, and Patient Safety* (pp. 197–222). Hershey, PA: IGI Global. DOI: 10.4018/978-1-5225-2337-6.ch008

Smith, C. R. (2018). Medical Students' Quest Towards the Long White Coat: Impact on Mental Health and Well-Being. In Smith, C. (Ed.), *Exploring the Pressures of Medical Education From a Mental Health and Wellness Perspective* (pp. 1–42). Hershey, PA: IGI Global. DOI: 10.4018/978-1-5225-2811-1.ch001

Smith, S. I., & Dandignac, M. (2018). Perfectionism: Addressing Lofty Expectations in Medical School. In Smith, C. (Ed.), *Exploring the Pressures of Medical Education From a Mental Health and Wellness Perspective* (pp. 68–91). Hershey, PA: IGI Global. DOI: 10.4018/978-1-5225-2811-1.ch003

Soczywko, J., & Rutkowska, D. (2018). The Patient/Provider Relationship in Emergency Medicine: Organization, Communication, and Understanding. In Rosiek-Kryszewska, A., & Leksowski, K. (Eds.), *Healthcare Administration for Patient Safety and Engagement* (pp. 74–105). Hershey, PA: IGI Global. DOI: 10.4018/978-1-5225-3946-9.ch005

Soni, P. (2018). Implications of HIPAA and Subsequent Regulations on Information Technology. In Gupta, M., Sharman, R., Walp, J., & Mulgund, P. (Eds.), *Information Technology Risk Management and Compliance in Modern Organizations* (pp. 71–98). Hershey, PA: IGI Global. DOI: 10.4018/978-1-5225-2604-9.ch004

Souza, T. de V. I. E. (2020). Certifications for Medical Interpreters: A Comparative Analysis. In I. Souza, & E. Fragkou (Eds.), *Handbook of Research on Medical Interpreting* (pp. 26-53). IGI Global. https://doi.org/10.4018/978-1-5225-9308-9.ch002

Srivastava, A., & Aggarwal, A. K. (2018). Medical Image Fusion in Spatial and Transform Domain: A Comparative Analysis. In Anwar, M., Khosla, A., & Kapoor, R. (Eds.), *Handbook of Research on Advanced Concepts in Real-Time Image and Video Processing* (pp. 281–300). Hershey, PA: IGI Global. DOI: 10.4018/978-1-5225-2848-7.ch011

Srivastava, S. K., & Roy, S. N. (2018). Recommendation System: A Potential Tool for Achieving Pervasive Health Care. In Machado, J., Abelha, A., Santos, M., & Portela, F. (Eds.), *Next-Generation Mobile and Pervasive Healthcare Solutions* (pp. 111–127). Hershey, PA: IGI Global. DOI: 10.4018/978-1-5225-2851-7.ch008

Stanimirovic, D. (2017). Digitalization of Death Certification Model: Transformation Issues and Implementation Concerns. In Saeed, S., Bamarouf, Y., Ramayah, T., & Iqbal, S. (Eds.), *Design Solutions for User-Centric Information Systems* (pp. 22–43). Hershey, PA: IGI Global. DOI: 10.4018/978-1-5225-1944-7.ch002

Subbaraman, K., Singh, M., & Johar, I. P. (2021). Medical Tourism: History, Global Scenario, and Indian Perspectives. In Singh, M., & Kumaran, S. (Eds.), *Growth of the Medical Tourism Industry and Its Impact on Society: Emerging Research and Opportunities* (pp. 1–18). IGI Global. https://doi.org/10.4018/978-1-7998-3427-4.ch001

Sukkird, V., & Shirahada, K. (2018). E-Health Service Model for Asian Developing Countries: A Case of Emergency Medical Service for Elderly People in Thailand. In Khosrow-Pour, M. (Ed.), *Optimizing Current Practices in E-Services and Mobile Applications* (pp. 214–232). Hershey, PA: IGI Global. DOI: 10.4018/978-1-5225-5026-6.ch011

Talbot, T. B. (2017). Making Lifelike Medical Games in the Age of Virtual Reality: An Update on "Playing Games with Biology" from 2013. In Dubbels, B. (Ed.), *Transforming Gaming and Computer Simulation Technologies across Industries* (pp. 103–119). Hershey, PA: IGI Global. DOI: 10.4018/978-1-5225-1817-4.ch006

Tarver, E. (2022). Virtual Simulation: A Flipped Classroom Teaching Tool for Healthcare Education. In Coelho, L., Queirós, R., & Reis, S. (Eds.), *Emerging Advancements for Virtual and Augmented Reality in Healthcare* (pp. 65–81). IGI Global. https://doi.org/10.4018/978-1-7998-8371-5.ch005

Tripathi, S., & Musiolik, T. H. (2022). Fairness and Ethics in Artificial Intelligence-Based Medical Imaging. In Musiolik, T., & Dingli, A. (Eds.), *Ethical Implications of Reshaping Healthcare With Emerging Technologies* (pp. 71–85). IGI Global. https://doi.org/10.4018/978-1-7998-7888-9.ch004

Turcu, C. E., & Turcu, C. O. (2017). Social Internet of Things in Healthcare: From Things to Social Things in Internet of Things. In Reis, C., & Maximiano, M. (Eds.), *Internet of Things and Advanced Application in Healthcare* (pp. 266–295). Hershey, PA: IGI Global. DOI: 10.4018/978-1-5225-1820-4.ch010

Tutgun Ünal, A., Ekinci, Y., & Tarhan, N. (2022). Health Literacy and Cyberchondria. In Aker, H., & Aiken, M. (Eds.), *Handbook of Research on Cyberchondria, Health Literacy, and the Role of Media in Society's Perception of Medical Information* (pp. 276–297). IGI Global. https://doi.org/10.4018/978-1-7998-8630-3.ch015

Unwin, D. W., Sanzogni, L., & Sandhu, K. (2017). Developing and Measuring the Business Case for Health Information Technology. In Moahi, K., Bwalya, K., & Sebina, P. (Eds.), *Health Information Systems and the Advancement of Medical Practice in Developing Countries* (pp. 262–290). Hershey, PA: IGI Global. DOI: 10.4018/978-1-5225-2262-1.ch015

Ustun, A. B., Yilmaz, R., & Yilmaz, F. G. (2020). Virtual Reality in Medical Education. In Umair, S. (Ed.), *Mobile Devices and Smart Gadgets in Medical Sciences* (pp. 56–73). IGI Global. https://doi.org/10.4018/978-1-7998-2521-0.ch004

Vamsi, D., & Reddy, P. (2020). Electronic Health Record Security in Cloud: Medical Data Protection Using Homomorphic Encryption Schemes. In Chakraborty, C. (Ed.), *Smart Medical Data Sensing and IoT Systems Design in Healthcare* (pp. 22–47). IGI Global. https://doi.org/10.4018/978-1-7998-0261-7.ch002

Vartholomaios, P., Ramdani, N., Christophorou, C., Georgiadis, D., Guilcher, T., Blouin, M., Rebiai, M., Panayides, A. S., Pattichis, C. S., Sarafidis, M., Costarides, V., Vellidou, E., & Koutsouris, D. (2019). ENDORSE Concept: An Integrated Indoor Mobile Robotic System for Medical Diagnostic Support. *International Journal of Reliable and Quality E-Healthcare*, 8(3), 47–59. https://doi.org/10.4018/IJRQEH.2019070104

Vasant, P. (2018). A General Medical Diagnosis System Formed by Artificial Neural Networks and Swarm Intelligence Techniques. In Kose, U., Guraksin, G., & Deperlioglu, O. (Eds.), *Nature-Inspired Intelligent Techniques for Solving Biomedical Engineering Problems* (pp. 130–145). Hershey, PA: IGI Global. DOI: 10.4018/978-1-5225-4769-3.ch006

Vaz de Almeida, C. (2022). ACP Model-Assertiveness, Clarity, and Positivity: A Communication and Health Literacy Model for Health Professionals. In Vaz de Almeida, C., & Ramos, S. (Eds.), *Handbook of Research on Assertiveness, Clarity, and Positivity in Health Literacy* (pp. 1–22). IGI Global. https://doi.org/10.4018/978-1-7998-8824-6.ch001

Verma, K. (2022). Modeling Digital Healthcare Services Using NLP and IoT in Smart Cities. In Thomas, J., Geropanta, V., Karagianni, A., Panchenko, V., & Vasant, P. (Eds.), *Smart Cities and Machine Learning in Urban Health* (pp. 138–155). IGI Global. https://doi.org/10.4018/978-1-7998-7176-7.ch007

Viswanathan, J., Saranya, N., & Inbamani, A. (2021). Deep Learning Applications in Medical Imaging: Introduction to Deep Learning-Based Intelligent Systems for Medical Applications. In Saxena, S., & Paul, S. (Eds.), *Deep Learning Applications in Medical Imaging* (pp. 156–177). IGI Global. https://doi.org/10.4018/978-1-7998-5071-7.ch007

Wang, X., Su, K., & Su, L. (2019). Research on Improved Apriori Algorithm Based on Data Mining in Electronic Cases. *International Journal of Healthcare Information Systems and Informatics*, 14(3), 16–28. https://doi.org/10.4018/IJHISI.2019070102

Wickramasinghe, N. (2020). Minority Health and Wellness: A Digital Health Opportunity. In Wickramasinghe, N. (Ed.), *Handbook of Research on Optimizing Healthcare Management Techniques* (pp. 116–126). IGI Global. https://doi.org/10.4018/978-1-7998-1371-2.ch008

Wickramasinghe, N., Geholt, V., Sloane, E., Smart, P. J., & Schaffer, J. L. (2021). Using Health 4.0 to Enable Post-Operative Wellness Monitoring: The Case of Colorectal Surgery. In Wickramasinghe, N. (Ed.), *Optimizing Health Monitoring Systems With Wireless Technology* (pp. 233–247). IGI Global. https://doi.org/10.4018/978-1-5225-6067-8.ch016

Wilcox, S., Huzo, O., Minhas, A., Walters, N., Adada, J. E., Pennington, M., Roseme, L., Mohammed, D., Dusic, A., & Zeine, R. (2022). The Impact of Medical or Health-Related Internet Searches on Patient Compliance: The Dr. Net Study. In Aker, H., & Aiken, M. (Eds.), *Handbook of Research on Cyberchondria, Health Literacy, and the Role of Media in Society's Perception of Medical Information* (pp. 72–97). IGI Global. https://doi.org/10.4018/978-1-7998-8630-3.ch005

Witzke, K., & Specht, O. (2017). M-Health Telemedicine and Telepresence in Oral and Maxillofacial Surgery: An Innovative Prehospital Healthcare Concept in Structurally Weak Areas. *International Journal of Reliable and Quality E-Healthcare*, 6(4), 37–48. DOI: 10.4018/IJRQEH.2017100105

Wong, A. K., & Lo, M. F. (2018). Using Pervasive Computing for Sustainable Healthcare in an Aging Population. In Fong, B., Ng, A., & Yuen, P. (Eds.), *Sustainable Health and Long-Term Care Solutions for an Aging Population* (pp. 187–202). Hershey, PA: IGI Global. DOI: 10.4018/978-1-5225-2633-9.ch010

Wu, W., Martin, B. C., & Ni, C. (2017). A Systematic Review of Competency-Based Education Effort in the Health Professions: Seeking Order Out of Chaos. In Rasmussen, K., Northrup, P., & Colson, R. (Eds.), *Handbook of Research on Competency-Based Education in University Settings* (pp. 352–378). Hershey, PA: IGI Global. DOI: 10.4018/978-1-5225-0932-5.ch018

Yadav, S., Ekbal, A., Saha, S., Pathak, P. S., & Bhattacharyya, P. (2017). Patient Data De-Identification: A Conditional Random-Field-Based Supervised Approach. In S. Saha, A. Mandal, A. Narasimhamurthy, S. V, & S. Sangam (Eds.), *Handbook of Research on Applied Cybernetics and Systems Science* (pp. 234-253). Hershey, PA: IGI Global. DOI: 10.4018/978-1-5225-2498-4.ch011

Yu, B., Wijesekera, D., & Costa, P. C. (2017). Informed Consent in Healthcare: A Study Case of Genetic Services. In Moumtzoglou, A. (Ed.), *Design, Development, and Integration of Reliable Electronic Healthcare Platforms* (pp. 211–242). Hershey, PA: IGI Global. DOI: 10.4018/978-1-5225-1724-5.ch013

Yu, B., Wijesekera, D., & Costa, P. C. (2017). Informed Consent in Electronic Medical Record Systems. In I. Management Association (Ed.), *Healthcare Ethics and Training: Concepts, Methodologies, Tools, and Applications* (pp. 1029-1049). Hershey, PA: IGI Global. DOI: 10.4018/978-1-5225-2237-9.ch049

Yu, M., Li, J., & Wang, W. (2017). Creative Life Experience among Students in Medical Education. In Zhou, C. (Ed.), *Handbook of Research on Creative Problem-Solving Skill Development in Higher Education* (pp. 158–184). Hershey, PA: IGI Global. DOI: 10.4018/978-1-5225-0643-0.ch008

Zanetti, C. A., George, A., Stiegmann, R. A., & Phelan, D. (2020). Digital Health. In Gotian, R., Kang, Y., & Safdieh, J. (Eds.), *Handbook of Research on the Efficacy of Training Programs and Systems in Medical Education* (pp. 404–426). IGI Global. https://doi.org/10.4018/978-1-7998-1468-9.ch021

Zangão, M. O. (2020). Self-Perceived Health Status. In Fonseca, C., Lopes, M., Mendes, D., Mendes, F., & García-Alonso, J. (Eds.), *Handbook of Research on Health Systems and Organizations for an Aging Society* (pp. 1–11). IGI Global. DOI: 10.4018/978-1-5225-9818-3.ch001

Zarour, K. (2017). Towards a Telehomecare in Algeria: Case of Diabetes Measurement and Remote Monitoring. *International Journal of E-Health and Medical Communications*, 8(4), 61–80. DOI: 10.4018/IJEHMC.2017100104

Zavyalova, Y. V., Korzun, D. G., Meigal, A. Y., & Borodin, A. V. (2017). Towards the Development of Smart Spaces-Based Socio-Cyber-Medicine Systems. *International Journal of Embedded and Real-Time Communication Systems*, 8(1), 45–63. DOI: 10.4018/IJERTCS.2017010104

Zeinali, A. A. (2018). Word Formation Study in Developing Naming Guidelines in the Translation of English Medical Terms Into Persian. In M. Khosrow-Pour, D.B.A. (Ed.), *Encyclopedia of Information Science and Technology, Fourth Edition* (pp. 5136-5147). Hershey, PA: IGI Global. DOI: 10.4018/978-1-5225-2255-3.ch446

Zineldin, M., & Vasicheva, V. (2018). Reducing Medical Errors and Increasing Patient Safety: TRM and 5 Q's Approaches for Better Quality of Life. In *Technological Tools for Value-Based Sustainable Relationships in Health: Emerging Research and Opportunities* (pp. 87–115). Hershey, PA: IGI Global. DOI: 10.4018/978-1-5225-4091-5.ch005

About the Contributors

Yury Kovalyov, Doctor of Technical Sciences, Professor

Nver Mkhitaryan, Doctor of Technical Sciences, Professor

Andriy Morozov, Doctor of Philosophical Sciences, Professor

Yaroslava Zhukova, Ph.D (Biology)

Index

A

ancient 19, 68, 73, 78, 79, 80, 82, 83, 85, 86, 90, 92, 95, 97, 99, 100, 110, 113, 114, 121, 127, 131, 133, 134, 136, 145, 153, 155, 160, 161, 171, 174, 175, 176, 179, 193, 203, 211, 219, 222, 224, 234, 235, 243, 245, 249

Axiomatic 3, 4, 5, 6, 7, 8, 28

B

body 2, 5, 16, 17, 18, 22, 23, 24, 25, 26, 30, 33, 34, 35, 36, 37, 40, 41, 44, 46, 48, 49, 51, 52, 59, 62, 64, 65, 67, 78, 86, 88, 100, 101, 104, 105, 106, 107, 109, 110, 111, 112, 113, 114, 115, 116, 119, 120, 121, 122, 123, 124, 125, 130, 131, 132, 136, 142, 143, 145, 146, 147, 149, 153, 154, 155, 158, 159, 160, 162, 164, 165, 166, 167, 168, 169, 170, 171, 172, 173, 175, 179, 185, 186, 188, 190, 191, 192, 194, 196, 197, 198, 200, 202, 211, 229, 230

burial 81, 96, 130, 135, 136, 139, 140, 153, 167, 178, 179, 180

C

channels 20, 21, 22, 24, 26, 28, 35, 46, 48, 49, 52, 53, 55, 56, 57, 58, 68, 72, 74, 78, 79, 80, 81, 82, 84, 85, 86, 87, 88, 89, 91, 92, 99, 107, 109, 114, 115, 116, 121, 123, 124, 134, 144, 145, 152, 153, 155, 160, 161, 164, 198

Chapter 1, 2, 7, 16, 27, 28, 33, 34, 35, 37, 39, 44, 45, 49, 50, 53, 54, 55, 56, 57, 62, 67, 69, 70, 72, 73, 74, 75, 90, 91, 99, 100, 101, 113, 119, 120, 125, 126, 131, 132, 133, 134, 140, 142, 143, 144, 145, 146, 149, 151, 155, 157, 158, 159, 161, 162, 165, 171, 172, 174, 175, 183, 189, 192, 193, 198, 199, 203, 204, 205, 217, 221, 234

complex 1, 2, 4, 7, 9, 15, 20, 23, 24, 28, 30, 33, 39, 40, 45, 46, 48, 62, 65, 67, 81, 82, 84, 87, 96, 99, 102, 111, 122, 127, 133, 139, 140, 155, 174, 200, 224, 225

components 3, 5, 6, 7, 8, 14, 15, 16, 27, 31, 33, 34, 36, 37, 39, 40, 44, 45, 48, 60, 62, 101, 107, 110, 115, 120, 124, 155, 174, 200

connective 2, 28, 41, 68, 69, 70, 71, 72, 73, 75, 77, 78, 79, 80, 81, 84, 85, 86, 87, 90, 91, 92, 115, 127, 134, 142, 147, 155, 159, 160, 174

consciousness 10, 21, 22, 29, 30, 34, 35, 46, 47, 73, 79, 85, 87, 102, 112, 113, 114, 115, 121, 122, 123, 124, 125, 133, 134, 135, 142, 160, 162, 163, 174, 175, 176, 183, 189, 190, 191, 192, 194, 197, 200, 203, 221, 224, 225, 228, 230, 232, 237, 239, 241, 242, 244, 245

convolution 3, 15, 24, 28, 33, 34, 37, 44, 45, 46, 47, 52, 53, 55, 56, 57, 58, 59, 60, 62, 91, 92, 115, 120, 121, 127, 133, 147, 152, 154, 159, 160, 161, 162, 163, 164, 172, 190, 193, 194, 197, 198

D

data 2, 11, 16, 21, 24, 27, 33, 34, 39, 51, 62, 63, 67, 68, 69, 72, 77, 79, 85, 87, 88, 91, 92, 100, 107, 120, 124, 133, 134, 140, 142, 183, 217

dead 49, 53, 54, 55, 58, 60, 67, 68, 69, 75, 78, 79, 82, 83, 85, 86, 88, 89, 91, 92, 93, 94, 95, 96, 97, 98, 100, 101, 102, 103, 110, 113, 114, 120, 121, 122, 127, 128, 129, 131, 133, 134, 136, 137, 138, 139, 140, 142, 143, 144, 145, 146, 147, 153, 155, 156, 157, 160, 161, 163, 164, 165, 166, 167, 168, 172, 173, 174, 176, 179, 183, 185, 188, 191, 192, 196, 197, 198, 199, 200, 201, 203, 206, 207, 211, 212, 216, 217, 235, 236, 240, 246, 247, 248

death 24, 25, 27, 28, 33, 34, 35, 37, 39,

40, 41, 46, 48, 49, 50, 52, 53, 54, 55, 56, 57, 58, 61, 62, 64, 65, 67, 69, 74, 75, 83, 84, 86, 88, 91, 93, 94, 95, 98, 101, 104, 105, 107, 110, 112, 113, 115, 116, 119, 120, 121, 126, 128, 131, 132, 134, 138, 144, 145, 147, 148, 151, 152, 154, 158, 159, 161, 162, 163, 164, 165, 166, 167, 169, 170, 172, 173, 174, 180, 183, 184, 185, 186, 187, 188, 189, 190, 192, 196, 200, 201, 212, 213, 214, 215, 216, 217, 219, 221, 222, 223, 224, 225, 230, 231, 232, 233, 234, 235, 236, 237, 238, 239, 240, 241, 242, 243, 244, 245, 246, 247, 248, 249

destruction 25, 27, 31, 33, 36, 37, 39, 40, 45, 49, 52, 62, 67, 74, 110, 112, 120, 138, 164, 227, 235, 238

E

Elements 3, 4, 5, 6, 7, 8, 10, 12, 15, 17, 23, 31, 35, 43, 45, 83, 112, 121, 124, 131, 137, 157, 189, 232, 237

existence 1, 2, 3, 4, 5, 13, 27, 28, 31, 33, 34, 35, 39, 40, 44, 45, 46, 60, 62, 63, 65, 67, 68, 71, 72, 75, 77, 80, 81, 84, 86, 88, 89, 92, 99, 100, 101, 109, 110, 112, 114, 116, 120, 123, 124, 126, 127, 137, 138, 145, 159, 160, 174, 175, 185, 190, 198, 199, 200, 221, 223, 224, 225, 228, 229, 230, 231, 232, 235, 238, 242, 243, 244, 245

existing 3, 4, 6, 8, 10, 12, 24, 61, 127, 174, 227

experience 8, 10, 34, 46, 53, 55, 58, 67, 74, 86, 87, 99, 115, 126, 160, 174, 192, 199, 200, 221, 222, 223, 224, 226, 229, 231, 232, 233, 234, 238, 240, 241, 242, 243, 244, 245, 247, 249, 250

expressed 13, 14, 22, 23, 26, 46, 67, 70, 72, 89, 91, 138, 163, 174, 175, 200, 226

external 4, 5, 9, 10, 18, 20, 21, 22, 23, 24, 25, 26, 27, 35, 36, 37, 39, 40, 46, 49, 50, 51, 52, 54, 57, 58, 59, 61, 62, 113, 114, 125, 147, 161, 230, 233, 238

F

Fairy 68, 69, 72, 79, 81, 85, 88, 90, 97, 124, 179, 203, 204, 206, 211, 217, 219

G

Gilgamesh 68, 74, 75, 79, 84, 85, 91, 93, 103, 104, 106, 128, 129, 130, 143, 147, 148, 149, 150, 151, 152, 153, 154, 160, 162, 169, 171, 176, 178, 204, 209, 211

H

human 1, 2, 10, 15, 16, 17, 18, 20, 21, 23, 24, 27, 29, 30, 35, 41, 43, 44, 61, 62, 64, 65, 68, 71, 72, 73, 78, 84, 86, 94, 95, 96, 99, 100, 110, 112, 113, 114, 116, 119, 120, 121, 122, 125, 126, 129, 130, 133, 134, 136, 142, 146, 149, 155, 161, 163, 166, 169, 171, 174, 176, 178, 183, 185, 197, 200, 201, 203, 217, 221, 223, 224, 225, 228, 229, 230, 231, 232, 233, 235, 236, 237, 238, 239, 241, 243, 244, 248

I

information 67, 70, 72, 79, 88, 91, 92, 133, 200, 204, 216, 248

interaction 3, 5, 7, 8, 16, 22, 26, 31, 35, 47, 49, 50, 71, 122, 134, 153, 160

intuitive 1, 11, 12, 14, 15, 19, 26, 67, 68, 73, 74, 77, 80, 90, 92, 103, 113, 127, 175, 199, 200, 229, 234

J

Jehovah 116, 117, 118, 162

L

lords 69, 204, 210, 212, 213, 214, 215, 216

M

man 15, 23, 30, 35, 65, 68, 73, 85, 86, 93, 95, 104, 105, 107, 109, 112, 113, 114, 115, 116, 117, 118, 120, 123, 125, 135, 140, 142, 144, 149, 162, 163, 164, 169, 172, 173, 185, 186, 187, 199, 203, 207, 208, 209, 224, 225, 231, 232, 233, 236, 237, 238, 239, 240, 242, 243, 246

modeling 2, 4, 5, 13, 15, 24, 30, 34, 35, 61, 62, 63, 65, 175, 200, 217, 219

O

operation 5, 6, 7, 8, 9, 31

P

person 10, 17, 19, 21, 22, 23, 24, 25, 28, 35, 36, 37, 39, 40, 46, 48, 49, 53, 57, 58, 62, 64, 67, 70, 77, 80, 81, 82, 86, 87, 89, 91, 100, 101, 102, 106, 107, 109, 111, 112, 114, 115, 116, 118, 119, 121, 122, 123, 124, 125, 134, 136, 144, 152, 153, 163, 164, 169, 171, 184, 190, 193, 194, 195, 196, 221, 223, 224, 225, 227, 229, 230, 231, 232, 233, 234, 235, 237, 238, 239, 240, 241, 242, 243, 244, 245

phenomena 10, 15, 34, 39, 40, 41, 53, 58, 62, 68, 200, 228, 244, 245

place 16, 41, 45, 52, 54, 60, 68, 72, 75, 77, 91, 101, 126, 136, 139, 142, 147, 151, 153, 159, 167, 169, 171, 180, 186, 187, 194, 197, 198, 205, 210, 211, 215, 222, 226, 231, 233, 235, 237, 238, 241, 249

posthumous 1, 2, 27, 28, 41, 63, 67, 80, 81, 84, 86, 88, 89, 90, 100, 101, 102, 126, 134, 142, 143, 155, 159, 167, 172, 173, 174, 175, 183, 185, 188, 200, 201, 237

post-mortem 33, 40, 44, 59, 61, 63, 67, 90, 91, 99, 100, 105, 113, 125, 126, 127, 133, 134, 138, 147, 151, 152, 157, 158, 160, 165, 168, 171, 173, 174, 175, 183, 189, 190, 192, 194, 196, 197, 198, 200, 204, 211

properties 2, 6, 7, 14, 24, 26, 28, 31, 44, 45, 46, 49, 58, 60, 69, 71, 136, 151, 153, 194, 198

S

scenario 2, 9, 15, 16, 20, 21, 23, 24, 27, 28, 33, 34, 36, 39, 47, 59, 61, 67, 90, 91, 99, 101, 115, 119, 120, 121, 125, 126, 127, 133, 134, 137, 138, 147, 151, 152, 154, 155, 158, 159, 160, 164, 165, 167, 168, 171, 172, 173, 174, 175, 183, 190, 192, 193, 195, 196, 197, 198, 199, 200, 211

scenarios 1, 2, 15, 28, 33, 34, 35, 38, 44, 47, 61, 62, 63, 67, 90, 99, 100, 125, 126, 133, 134, 143, 157, 161, 173, 183, 189, 198, 199, 200, 204

self-organization 1, 2, 4, 15, 20, 21, 23, 28, 29, 33, 34, 35, 36, 39, 45, 46, 61, 62, 64, 67, 70, 72, 94, 129, 142, 147, 159, 164, 174, 176, 200, 201, 248

soliton 3, 5, 6, 7, 8, 16, 24, 25, 26, 27, 31, 33, 36, 37, 39, 40, 44, 45, 46, 50, 60, 61, 67, 99, 101, 102, 107, 109, 110, 114, 115, 120, 121, 124, 125, 138, 155, 156, 159, 164, 174, 198, 199, 200

soliton-wave 26, 33, 36, 37, 39, 40, 45, 60, 61, 67, 99, 101, 102, 109, 110, 114, 115, 120, 121, 124, 125, 138, 156, 159, 164, 174, 198, 200

Sources 2, 18, 27, 34, 52, 67, 68, 69, 72, 77, 79, 80, 87, 89, 90, 91, 92, 99, 100, 102, 125, 126, 133, 134, 174, 183, 198, 199, 200, 203, 222

Sp 2, 3, 4, 5, 7, 10, 11, 12, 15, 16, 24, 25, 31, 58, 70

spirit 20, 71, 78, 100, 104, 107, 109, 110, 112, 113, 114, 115, 116, 119, 120, 121, 124, 125, 130, 131, 134, 138, 158, 159, 160, 162, 164, 173, 179, 190, 192, 196, 230, 231, 233, 236, 238, 239, 240, 241, 244, 248

spiritual 111, 112, 113, 114, 115, 116, 119, 122, 124, 160, 169, 223, 224,

225, 226, 227, 229, 230, 231, 232, 233, 234, 235, 237, 238, 239, 240, 241, 242, 243, 244, 245, 247

S-space 1, 2, 3, 5, 7, 12, 15, 22, 28, 34, 35, 52, 65

statement 3, 4, 5, 6, 7, 8, 12, 13, 14, 15, 19, 226, 228, 234, 241

structure 1, 2, 3, 4, 5, 9, 16, 19, 21, 22, 23, 24, 28, 29, 46, 50, 57, 68, 70, 82, 89, 90, 110, 114, 115, 122, 125, 130, 137, 140, 178, 226, 231, 238

subjective 1, 2, 12, 16, 19, 20, 21, 22, 23, 24, 28, 34, 36, 45, 46, 47, 48, 49, 50, 52, 54, 58, 60, 62, 63, 67, 71, 91, 92, 99, 101, 102, 110, 115, 125, 127, 132, 144, 145, 155, 198

T

ternary 1, 2, 10, 21, 27, 28, 31, 49, 59, 67, 68, 69, 70, 71, 72, 73, 74, 75, 77, 78, 79, 80, 81, 84, 85, 86, 87, 89, 90, 91, 92, 113, 114, 115, 127, 134, 140, 142, 147, 155, 159, 160, 174, 175, 200, 204

Tibetan 68, 79, 88, 89, 93, 94, 95, 98, 100, 121, 128, 129, 131, 174, 176, 183, 184, 188, 197, 198, 199, 200, 201, 235, 246

U

underworld 69, 85, 87, 101, 104, 106, 115, 133, 140, 143, 144, 145, 146, 147, 154, 160, 171, 196, 203, 204, 207, 209, 210, 211, 212, 213, 218, 219

V

verification 14, 27, 28, 34, 46, 58, 61, 62, 67, 69, 72, 78, 90, 99, 126, 133, 174, 183, 202, 222

W

waves 3, 4, 5, 6, 7, 8, 17, 24, 31, 33, 50, 51, 52, 54, 60, 64, 153, 159, 189, 191